多语种工程机械型号名谱

英汉工程机械型号名谱

中国工程机械学会　编

李安虎　译

上海科学技术出版社

编　委　会

序

土石方工程、流动起重装卸工程、人货升降输送工程和各种建筑工程综合机械化施工以及同上述相关的工业生产过程的机械化作业所需的机械设备统称为工程机械。工程机械应用范围极广，大致涉及如下领域：① 交通运输基础设施；② 能源领域工程；③ 原材料领域工程；④ 农林基础设施；⑤ 水利工程；⑥ 城市工程；⑦ 环境保护工程；⑧ 国防工程。

工程机械行业的发展历程大致可分为以下 6 个阶段。

第一阶段(1949 年前)：工程机械最早应用于抗日战争时期滇缅公路建设。

第二阶段(1949—1960 年)：我国实施第一个和第二个五年计划，156 项工程建设需要大量工程机械，国内筹建了一批以维修为主、生产为辅的中小型工程机械企业，没有建立专业化的工程机械制造厂，没有统一的管理与规划，高等学校也未设立真正意义上的工程机械专业或学科，相关科研机构也没有建立。各主管部委虽然设立了一些管理机构，但这些机构分散且规模很小。此期间全行业的职工人数仅 2 万余人，生产企业仅二十余家，总产值 2.8 亿元人民币。

第三阶段(1961—1978 年)：国务院和中央军委决定在第一机械工业部成立工程机械工业局(五局)，并于 1961 年 4 月 24 日正式成立，由此对工程机械行业的发展进行统一规划，形成了独立的制造体系。此外，高等学校设立了工程机械专业以培养相应人才，并成立了独立的研究所以制定全行业的标准化和技术情报交流体系。在此期间，全行业职工人数达 34 万余人，全国工程机械专业厂和兼并厂达 380 多家，固定资产 35 亿元人民币，工业总产值 18.8 亿元人民币，毛利润 4.6 亿元人民币。

第四阶段(1979—1998 年)：这一时期工程机械管理机构经过几次大的变动，主要生产厂下放至各省、市、地区管理，改革开放的实行也促进了民营企业的发展。在此期间，全行业固定资产总额 210 亿元人民币，净值 140 亿元人民币，有 1 000 多家厂商，销售总额 350 亿元人民币。

第五阶段(1999—2012 年)：此阶段工程机械行业发展很快，成绩显著。全国有 1 400 多家厂商、主机厂 710 家，11 家企业入选世界工程机械 50 强，30 多家企业在 A 股和 H 股上市，销售总额已超过美国、德国、日本，位居世界第一，2012 年总产值近 5 000 亿元人民币。

第六阶段(2012 年至今)：在此期间国家进行了经济结构调整，工程机械行业的发展速度也有所变化，总体稳中有进。在经历了一段不景气的时期之后，随着我国“一带一路”倡议的实施和国内城乡建设的需要，将会迎来新的发展时期，完成由工程机械制造大国向工程机械制造强国的转变。

随着经济发展的需要，我国的工程机械行业逐渐发展壮大，由原来的以进口为主转向出口为主。1999 年至 2010 年期间，工程机械的进口额从 15.5 亿美元增长到 84 亿美元，而出口的变化更大，从 6.89 亿美元增长到 103.4 亿美元，2015 年达到近 200 亿美元。我国的工程机械已经出口到世界 200 多个国家和地区。

我国工程机械的品种越来越多，根据中国工程机械工业协会标准，我国工程机械已经形成 20 个大类、130 多个组、近 600 个型号、上千个产品，在这些产品中还不包括港口机械以及部分矿山机械。为了适应工程机械的出口需要和国内外行业的技术交流，我们将上述产品名称翻译成 8 种语言，包括阿拉伯语、德语、法语、日语、西班牙语、意大利语、英语和俄语，并分别提供中文对照，以方便大家在使用中进行参考。翻译如有不准确、不正确之处，恳请读者批评指正。

编委会

2020 年 1 月

目　　录

1 excavating machine 挖掘机械

Group/组	Type/型	Product/产品
intermittent excavator 间歇式挖掘机	mechanical excavator 机械式挖掘机	crawler mechanical excavator 履带式机械挖掘机
		wheel mechanical excavator 轮胎式机械挖掘机
		fixed (marine) mechanical excavator 固定式(船用)机械挖掘机
		mining shovel 矿用电铲
	hydraulic excavator 液压式挖掘机	crawler hydraulic excavator 履带式液压挖掘机
		wheel hydraulic excavator 轮胎式液压挖掘机
		amphibious hydraulic excavator 水陆两用式液压挖掘机
		wetland hydraulic excavator 湿地液压挖掘机
		walking hydraulic excavator 步履式液压挖掘机
		fixed (marine) hydraulic excavator 固定式(船用)液压挖掘机
	backhoe loader excavating loader loader-digger loader-excavator 挖掘装载机	side-shifting excavator loader 侧移式挖掘装载机
		intermediate excavator loader 中置式挖掘装载机
continuous excavator 连续式挖掘机	wheel bucket excavator 斗轮挖掘机	crawler wheel bucket excavator 履带式斗轮挖掘机
		wheel bucket excavator 轮胎式斗轮挖掘机
		special wheel bucket excavator 特殊行走装置斗轮挖掘机
	roll-cut excavator 滚切式挖掘机	roll-cut excavator 滚切式挖掘机
	milling excavator 铣切式挖掘机	milling excavator 铣切式挖掘机

（续表）

Group/组	Type/型	Product/产品
continuous excavator 连续式挖掘机	multi-bucket trencher 多斗挖沟机	forming section trencher 成型断面挖沟机
		bucket wheel trencher 轮斗挖沟机
		chain bucket trencher 链斗挖沟机
	chain bucket excavator 链斗挖沟机	crawler chain bucket excavator 履带式链斗挖沟机
		wheel chain bucket excavator 轮胎式链斗挖沟机
		track chain bucket excavator 轨道式链斗挖沟机
other excavation machinery 其他挖掘机械		

2 earthmoving machinery 铲土运输机械

Group/组	Type/型	Product/产品
loader 装载机	crawler loader 履带式装载机	mechanical loader 机械装载机
		hydraulic mechanical loader 液力机械装载机
		full hydraulic loader 全液压装载机
	wheel loader 轮胎式装载机	mechanical loader 机械装载机
		hydraulic mechanical loader 液力机械装载机
		full hydraulic loader 全液压装载机
	skid steering loader 滑移转向式装载机	skid steering loader 滑移转向装载机
	special purpose loader 特殊用途装载机	crawler wetland loader 履带湿地式装载机
		side loader 侧卸装载机

(续表)

Group/组	Type/型	Product/产品
loader 装载机	special purpose loader 特殊用途装载机	underground loader 井下装载机
		wood loader 木材装载机
carrier scraper, diesel scooptram, earth scraper, excavation scraper, load-haul-dump vehicle, power scraper, scoopmobile, scooptram, scraper 铲运机	self-propelled scraper 自行铲运机	self-propelled tyre scraper 自行轮胎式铲运机
		twin engine tyre scraper 轮胎式双发动机铲运机
		automatic crawler scraper 自行履带式铲运机
	tractor scraper 拖式铲运机	mechanical scraper 机械铲运机
		hydraulic scraper 液压铲运机
bulldozer, dozer, earth mover, push dozer, tractor dozer 推土机	crawler bulldozer 履带式推土机	mechanical bulldozer 机械推土机
		hydraulic mechanical bulldozer 液力机械推土机
		full hydraulic bulldozer 全液压推土机
		crawler wetland bulldozer 履带式湿地推土机
	wheel bulldozer 轮胎式推土机	hydraulic mechanical bulldozer 液力机械推土机
		full hydraulic bulldozer 全液压推土机
	tractor hoist 通井机	tractor hoist 通井机
	trimming dozer 推耙机	trimming dozer 推耙机
fork loader 叉装机	fork loader 叉装机	fork loader 叉装机
grader, plough, plow 平地机	self-propelled grader 自行式平地机	mechanical grader 机械式平地机
		hydraulic mechanical grader 液力机械平地机
		full hydraulic grader 全液压平地机

（续表）

Group/组	Type/型	Product/产品
grader，plough，plow 平地机	drawn grader，towed grader 拖式平地机	drawn grader，towed grader 拖式平地机
off-highway dump truck 非公路自卸车	rigid dump truck 刚性自卸车	mechanical drive dump truck 机械传动自卸车
		hydraulic mechanical drive dump truck 液力机械传动自卸车
		hydrostatic drive dump truck 静液压传动自卸车
		electric drive dump truck 电动自卸车
off-highway dump 非公路自卸车	articulated dump truck 铰接式自卸车	mechanical drive dump truck 机械传动自卸车
		hydraulic mechanical drive dump truck 液力机械传动自卸车
		hydrostatic drive dump truck 静液压传动自卸车
		electric drive dump truck 电动自卸车
	underground rigid dump truck 地下刚性自卸车	hydraulic mechanical drive dump truck 液力机械传动自卸车
	underground articulated dump truck 地下铰接式自卸车	hydraulic mechanical drive dump truck 液力机械传动自卸车
		hydrostatic drive dump truck 静液压传动自卸车
		electric drive dump truck 电动自卸车
	rotary dump truck 回转式自卸车	hydrostatic drive dump truck 静液压传动自卸车
	gravity dump truck 重力翻斗车	gravity dump truck 重力翻斗车
operation preparation machinery 作业准备机械	bush cutter 除荆机	bush cutter 除荆机
	stumper 除根机	stumper 除根机

（续表）

Group/组	Type/型	Product/产品
other earthmoving machinery 其他铲土运输机械		

3 hoisting machinery 起重机械

Group/组	Type/型	Product/产品
mobile crane 流动式起重机	rubber tyred crane, tyre crane, tyre-mounted crane, wheel crane 轮胎式起重机	mobile crane, truck crane, truck mounted crane 汽车起重机
		all terrain crane 全地面起重机
		tyre-mounted crane 轮胎式起重机
		tyre-mounted rough terrain crane 越野轮胎起重机
		lorry crane, lorry loading crane, lorry-mounted crane, vehicle mounted crane 随车起重机
	caterpillar crane, crawler crane 履带式起重机	crawler crane with trussed jib 桁架臂履带起重机
		telescopic crawler crane 伸缩臂履带起重机
	special mobile crane 专用流动式起重机	reach stacker 正面吊运起重机
		side-shifting crane 侧面吊运起重机
		crawler type pipe handler 履带式吊管机
	wrecker 清障车	wrecker 清障车
		wrecker 清障抢救车
building crane 建筑起重机械	tower crane 塔式起重机	orbit upper slewing tower crane 轨道上回转塔式起重机

（续表）

Group/组	Type/型	Product/产品
building crane 建筑起重机械	tower crane 塔式起重机	orbit upper slewing self-lifting tower crane 轨道上回转自升塔式起重机
		orbit lower slewing tower crane 轨道下回转塔式起重机
		orbit quick-mounted tower crane 轨道快装式塔式起重机
		orbit whip-typed tower crane 轨道动臂式塔式起重机
		orbit flat top tower crane 轨道平头式塔式起重机
		fixed upper slewing tower crane 固定上回转塔式起重机
		fixed upper slewing self-lifting tower crane 固定上回转自升塔式起重机
		fixed lower slewing tower crane 固定下回转塔式起重机
		fixed quick-mounted tower crane 固定快装式塔式起重机
	tower crane 塔式起重机	fixed whip-typed tower crane 固定动臂式塔式起重机
		fixed flat top tower crane 固定平头式塔式起重机
		fixed interior climbing tower crane 固定内爬升式塔式起重机
	construction hoist 施工升降机	gear and rack construction hoist 齿轮齿条式施工升降机
		rope suspended hoist 钢丝绳式施工升降机
		combined hoist 混合式施工升降机
	construction winch 建筑卷扬机	single-cylinder fast speed hoister 单筒卷扬机
		double-drum fast speed hoister 双筒式卷扬机
		tri-drum hoister 三筒式卷扬机
other hoisting machinery 其他起重机械		

4 industrial vehicle 工业车辆

Group/组	Type/型	Product/产品
power-driven industrial vehicle 机动工业车辆(内燃、蓄电池、双动力)	fixed platform carrier 固定平台搬运车	fixed platform carrier 固定平台搬运车
	towing car and pushing tractor 牵引车和推顶车	towing car 牵引车
		pushing tractor 推顶车
	stacking (high lift) vehicle 堆垛用(高起升)车辆	counter balanced forklift truck, counter balanced fork lift truck 平衡重式叉车
		movable type fork lift truck, reach truck 前移式叉车
		straddle arm type forklift truck, straddle truck, straddle-leg forklift truck 插腿式叉车
		pallet stacker, pallet-stacking truck 托盘堆垛车
		platform stacker 平台堆垛车
		truck with elevatable operation position 操作台可升降车辆
		side loading forklift truck 侧面式叉车(单侧)
		rough terrain forklift 越野叉车
		lateral stacking forklift 侧面堆垛式叉车(两侧)
		three-way stacking forklift 三向堆垛式叉车
		stacking high-lift straddle carrier 堆垛用高起升跨车
		balanced heavy container stacker 平衡重式集装箱堆高机

（续表）

Group/组	Type/型	Product/产品
power-driven industrial vehicle 机动工业车辆（内燃、蓄电池、双动力）	non-stacking（low-lift）vehicle 非堆垛用（低起升）车辆	pallet carrier 托盘搬运车
		platform carrier 平台搬运车
		no-stacking low-lift straddle carrier 非堆垛用低起升跨车
	telescopic forklift truck 伸缩臂式叉车	telescopic forklift truck 伸缩臂式叉车
		rough terrain telescopic forklift truck 越野伸缩臂式叉车
	order picking truck，order-picking vehicle 拣选车	order picking truck，order-picking vehicle 拣选车
	driverless vehicle 无人驾驶车辆	driverless vehicle 无人驾驶车辆
non-motorized industrial vehicle，non-motor industrial vehicle 非机动工业车辆	walking stacker 步行式堆垛车	walking stacker 步行式堆垛车
	walking pallet stacker 步行式托盘堆垛车	walking pallet stacker 步行式托盘堆垛车
	walking pallet carrier 步行式托盘搬运车	walking pallet carrier 步行式托盘搬运车
	walking scissor lift pallet carrier 步行剪叉式升降托盘搬运车	walking scissor lift pallet carrier 步行剪叉式升降托盘搬运车
other industrial vehicle 其他工业车辆		

5 compact machine，compacting equipment，compaction machinery，compactor 压实机械

Group/组	Type/型	Product/产品
static road roller 静作用压路机	towed road roller 拖式压路机	towed smooth drum road roller 拖式光轮压路机
	self-propelled road roller 自行式压路机	double-drum road roller 两轮光轮压路机

(续表)

Group/组	Type/型	Product/产品
static road roller 静作用压路机	self-propelled road roller 自行式压路机	double-drum hinged road roller 两轮铰接光轮压路机
		three-drum road roller 三轮光轮压路机
		three-drum hinged road roller 三轮铰接光轮压路机
vibration roller 振动压路机	smooth drum road roller, smooth wheel roller 光轮式压路机	double drums vibratory roller, double-drum tandem vibratory road roller, tandem vibratory roller 两轮串联振动压路机
		double-drum hinged vibratory road roller 两轮铰接振动压路机
		four-drum vibratory roller, four-wheel vibratory road roller 四轮振动压路机
	tire driving road roller 轮胎驱动式压路机	tire driving smooth-wheel vibratory road roller 轮胎驱动光轮振动压路机
		tire driving pad foot vibratory road roller 轮胎驱动凸块振动压路机
	towed roller 拖式压路机	towed vibratory road roller 拖式振动压路机
		towed pad foot vibratory road roller 拖式凸块振动压路机
	walk behind road roller 手扶式压路机	walk behind smooth-wheel vibratory road roller 手扶光轮振动压路机
		walk behind pad foot vibratory road roller 手扶凸块振动压路机
		walk behind vibratory road roller with steering mechanism 手扶带转向机构振动压路机
oscillation road roller 振荡压路机	smooth drum road roller 光轮式压路机	double drums vibratory roller, double-drum tandem vibratory road roller 两轮串联振荡压路机

(续表)

Group/组	Type/型	Product/产品
oscillation road roller 振荡压路机	smooth drum road roller 光轮式压路机	double-drum hinged vibratory road roller 两轮铰接振荡压路机
	tire driving road roller 轮胎驱动式压路机	tire driving smooth drum vibratory road roller 轮胎驱动式光轮振荡压路机
tire compactor, tyre roller, tyred roller, tire road roller, tyre roller 轮胎压路机	self-propelled road roller 自行式压路机	tire roller 轮胎压路机
		hinged tire roller 铰接式轮胎压路机
impact road roller, impact roller 冲击压路机	towed roller 拖式压路机	towed impact roller 拖式冲击压路机
	self-propelled road roller 自行式压路机	self-propelled impact road roller 自行式冲击压路机
combined road roller 组合式压路机	vibratory combined tire road roller 振动轮胎组合式压路机	vibratory combined tire road roller 振动轮胎组合式压路机
	vibratory oscillation road roller 振动振荡式压路机	vibratory oscillation road roller 振动振荡式压路机
vibration plate compactor 振动平板夯	electric plate compactor 电动式平板夯	electric vibration plate compactor 电动振动平板夯
	diesel plate compactor 内燃式平板夯	diesel vibration plate compactor 内燃振动平板夯
vibration impacting rammer 振动冲击夯	electric impacting rammer 电动式冲击夯	electric vibration impacting rammer 电动振动冲击夯
	internal combustion impacting rammer 内燃冲击夯	internal combustion vibration impacting rammer 内燃振动冲击夯
explosion rammer 爆炸式夯实机	explosion rammer 爆炸式夯实机	explosion rammer 爆炸式夯实机
frog-type rammer 蛙式夯实机	frog-type rammer 蛙式夯实机	frog-type rammer 蛙式夯实机

（续表）

Group/组	Type/型	Product/产品
landfill roller 垃圾填埋压实机	static road roller 静碾式压实机	static landfill roller 静碾式垃圾填埋压实机
	vibratory roller 振动式压实机	vibratory landfill roller 振动式垃圾填埋压实机
other compact machine 其他压实机械		

6 pavement construction and protect machinery 路面施工与养护机械

Group/组	Type/型	Product/产品
asphalt pavement construction machinery 沥青路面施工机械	asphalt batch mixer, asphalt concrete mixing plant, asphalt mixer, asphalt mixing plant 沥青混合料搅拌设备	intermittent asphalt mixing plant 强制间歇式沥青搅拌设备
		continuous asphalt mixing plant 强制连续式沥青搅拌设备
		roller continuous asphalt mixing plant 滚筒连续式沥青搅拌设备
		continuous asphalt mixing plant with double barrels 双滚筒连续式沥青搅拌设备
		intermittent asphalt mixing plant with double barrels 双滚筒间歇式沥青搅拌设备
		mobile asphalt mixing plant 移动式沥青搅拌设备
		container asphalt mixing equipment 集装箱式沥青搅拌设备
		environmental protection asphalt mixing equipment 环保型沥青搅拌设备
	asphalt paver 沥青混合料摊铺机	mechanical drive crawler asphalt paver 机械传动履带式沥青摊铺机
		hydraulic crawler asphalt paver 全液压履带式沥青摊铺机
		mechanical drive tyre asphalt paver 机械传动轮胎式沥青摊铺机

（续表）

Group/组	Type/型	Product/产品
asphalt pavement construction machinery 沥青路面施工机械	asphalt paver 沥青混合料摊铺机	full hydraulic tyre asphalt paver 全液压轮胎式沥青摊铺机
		double-decked asphalt paver 双层沥青摊铺机
		asphalt paver with spraying device 带喷洒装置沥青摊铺机
		roadside paver 路沿摊铺机
	asphalt mixture transporter 沥青混合料转运机	direct asphalt transporter 直传式沥青转运料机
		asphalt transporter with silo 带料仓式沥青转运料机
	asphalt distributor 沥青洒布机(车)	mechanical drive asphalt distributor 机械传动沥青洒布机(车)
		hydraulic drive asphalt distributor 液压传动沥青洒布机(车)
		pneumatic asphalt distributor 气压沥青洒布机
	chipping spreader, road grit machine, stone chip spreader 碎石撒布机	single conveyor belt chipping spreader 单输送带石屑撒布机
		double conveyor belt chipping spreader 双输送带石屑撒布机
		suspension simple stone chipping spreader 悬挂式简易石屑撒布机
		black gravel spreader 黑色碎石撒布机
	bituminous binders dispenser, liquid asphalt truck 液态沥青运输机	insulated asphalt tanker 保温沥青运输罐车
		semi-trailer insulated asphalt tanker 半拖挂保温沥青运输罐车
		simple truck-mounted asphalt tanker 简易车载式沥青罐车
	asphalt pump, binder pump, binder transfer unit, pump for hot bituminous binders 沥青泵	gear asphalt pump 齿轮式沥青泵
		plunger asphalt pump 柱塞式沥青泵
		screw asphalt pump 螺杆式沥青泵

(续表)

Group/组	Type/型	Product/产品
asphalt pavement construction machinery 沥青路面施工机械	asphalt valve 沥青阀	insulation three-way asphalt valve (manual, electric, pneumatic) 保温三通沥青阀(分手动、电动、气动)
		insulation two-way asphalt valve (manual, electric, pneumatic) 保温二通沥青阀(分手动、电动、气动)
		insulation two-way asphalt ball valve 保温二通沥青球阀
	asphalt tank 沥青贮罐	vertical asphalt tank 立式沥青贮罐
		horizontal asphalt tank 卧式沥青贮罐
		asphalt storage (station) 沥青库(站)
asphalt pavement construction machinery 沥青路面施工机械	asphalt heating and melting equipment 沥青加热熔化设备	flame heating fixed asphalt melting equipment 火焰加热固定式沥青熔化设备
		flame heating mobile asphalt melting equipment 火焰加热移动式沥青熔化设备
		steam heated fixed asphalt melting equipmen 蒸汽加热固定式沥青熔化设备
		steam heated mobile asphalt melting equipment 蒸汽加热移动式沥青熔化设备
		heat conductive oil heated fixed asphalt melting equipment 导热油加热固定式沥青熔化设备
		electric heated fixed asphalt melting equipment 电加热固定式沥青熔化设备
		electric heated mobile asphalt melting equipment 电加热移动式沥青熔化设备
		infrared heated fixed asphalt melting equipment 红外线加热固定式沥青熔化设备

(续表)

Group/组	Type/型	Product/产品
asphalt pavement construction machinery 沥青路面施工机械	asphalt heating and melting equipment 沥青加热熔化设备	infrared heated mobile asphalt melting equipment 红外线加热移动式沥青熔化设备
		solar heated fixed asphalt melting equipment 太阳能加热固定式沥青熔化设备
		solar heated mobile asphalt melting equipment 太阳能加热移动式沥青熔化设备
	asphalt filling equipment 沥青灌装设备	barreled asphalt filling equipment 筒装沥青灌装设备
		bagged asphalt filling equipment 袋装沥青灌装设备
	asphalt de-barrel device 沥青脱桶装置	fixed asphalt de-barrel device 固定式沥青脱桶装置
		mobile asphalt de-barrel device 移动式沥青脱桶装置
	asphalt modification equipment 沥青改性设备	mixing asphalt modification equipment 搅拌式沥青改性设备
		colloid mill asphalt modification equipment 胶体磨式沥青改性设备
	asphalt emulsion equipment, bituminous emulsions plant 沥青乳化设备	mobile asphalt emulsion equipment 移动式沥青乳化设备
		fixed asphalt emulsion equipment 固定式沥青乳化设备
concrete pavement construction machinery 水泥面施工机械	concrete paver 水泥混凝土摊铺机	slipform cement concrete paver 滑模式水泥混凝土摊铺机
		track cement concrete paver 轨道式水泥混凝土摊铺机
	multifunctional curb paving machine 多功能路缘石铺筑机	crawler cement concrete curb paving machine 履带式水泥混凝土路缘铺筑机
		orbital cement concrete curb paving machine 轨道式水泥混凝土路缘铺筑机
		tire cement concrete curb paving machine 轮胎式水泥混凝土路缘铺筑机

(续表)

Group/组	Type/型	Product/产品
concrete pavement construction machinery 水泥面施工机械	concrete pavement expansion joints cutter，concrete pavement grooving machine 切缝机	walk behind concrete pavement grooving machine 手扶式水泥混凝土路面切缝机
		orbital concrete pavement grooving machine 轨道式水泥混凝土路面切缝机
		tire concrete pavement grooving machine 轮胎式水泥混凝土路面切缝机
	concrete pavement vibrating beam 水泥混凝土路面振动梁	single beam cement concrete pavement vibrating beam 单梁式水泥混凝土路面振动梁
		double beam cement concrete pavement vibrating beam 双梁式水泥混凝土路面振动梁
	concrete pavement trowelling machine 水泥混凝土路面抹光机	electric concrete pavement trowelling machine 电动式水泥混凝土路面抹光机
		internal combustion concrete pavement trowelling machine 内燃式水泥混凝土路面抹光机
	dehydration device of concrete pavement 水泥混凝土路面脱水装置	vacuum dehydration device of concrete pavement 真空式水泥混凝土路面脱水装置
		air cushion membrane dehydration device of concrete pavement 气垫膜式水泥混凝土路面脱水装置
	cement concrete side trench paver 水泥混凝土边沟铺筑机	crawler cement concrete side trench paver 履带式水泥混凝土边沟铺筑机
		track cement concrete side ditch paver 轨道式水泥混凝土边沟铺筑机
		tire cement concrete side ditch paver 轮胎式水泥混凝土边沟铺筑机
	crack sealing machine 路面灌缝机	drag-type crack sealing machine 拖式路面灌缝机
		self-propelled crack sealing machine 自行式路面灌缝机

（续表）

Group/组	Type/型	Product/产品
pavement base construction machinery 路面基层施工机械	stabilized soil mixer 稳定土拌和机	crawler stabilized soil mixer 履带式稳定土拌和机
		tire stabilized soil mixer 轮胎式稳定土拌和机
	stabilized soil mixing equipment 稳定土拌和设备	forced stabilized soil mixing equipment 强制式稳定土拌和设备
		self-falling stabilized soil mixing equipment 自落式稳定土拌和设备
	stabilized soil spreader，paver 稳定土摊铺机	crawler stabilized soil spreader，paver 履带式稳定土摊铺机
		tire stabilized soil spreader，paver 轮胎式稳定土摊铺机
construction machinery for pavement accessory facilities 路面附属设施施工机械	guardrail construction machinery 护栏施工机械	pile-driver/pile-drawing machine 打桩、拔桩机
		pile-driving machine with rig 钻孔吊桩机
	line marking construction machinery 标线标志施工机械	normal temperature paint marking sprayer 常温漆标线喷涂机
		hot melt paint marking machine 热熔漆标线划线机
		marking cleaner 标线清除机
	side ditch and slope protection construction machinery 边沟、护坡施工机械	ditching machine 开沟机
		side ditch paver 边沟摊铺机
		slope paver 护坡摊铺机
pavement maintenance machinery 路面养护机械	multifunctional maintenance machine 多功能养护机	multifunctional maintenance machine 多功能养护机
	asphalt pavement pothole patcher 沥青路面坑槽修补机	asphalt pavement pothole patcher 沥青路面坑槽修补机

（续表）

Group/组	Type/型	Product/产品
pavement maintenance machinery 路面养护机械	asphalt pavement heating patcher 沥青路面加热修补机	asphalt pavement heating patcher 沥青路面加热修补机
	pothole spray patcher 喷射式坑槽修补机	pothole spray patcher 喷射式坑槽修补机
	regenerative repair machine 再生修补机	regenerative repair machine 再生修补机
	cracks machine 扩缝机	cracks machine 扩缝机
	trench trimmer 坑槽切边机	trench trimmer 坑槽切边机
	small cover machine 小型罩面机	small cover machine 小型罩面机
	concrete pavement cutters 路面切割机	concrete pavement cutter 路面切割机
	motor sprinkler，sprinkler，sprinkler truck，street sprinkler，water dispenser，watering cart 洒水车	motor sprinkler，sprinkler，sprinkler truck，street sprinkler，water dispenser，watering cart 洒水车
	pavement milling machine 路面刨铣机	crawler pavement milling machine，crawler pavement miller 履带式路面刨铣机
		tire pavement milling machine 轮胎式路面刨铣机
	asphalt pavement maintenance truck 沥青路面养护车	self-propelled asphalt pavement maintenance truck 自行式沥青路面养护车
		towed asphalt pavement maintenance truck 拖式沥青路面养护车
	cement concrete pavement maintenance truck 水泥混凝土路面养护车	self-propelled cement concrete pavement maintenance truck 自行式水泥混凝土路面养护车
		towed cement concrete pavement maintenance truck 拖式水泥混凝土路面养护车

（续表）

Group/组	Type/型	Product/产品
pavement maintenance machinery 路面养护机械	concrete crusher 水泥混凝土路面破碎机	self-propelled concrete crusher 自行式水泥混凝土路面破碎机
		towed concrete crusher 拖式水泥混凝土路面破碎机
	slurry sealing machine 稀浆封层机	self-propelled slurry sealing machine 自行式稀浆封层机
		towed slurry sealing machine 拖式稀浆封层机
	sand return machine 回砂机	scraper type sand return machine 刮板式回砂机
		rotor sand return machine 转子式回砂机
pavement maintenance machinery 路面养护机械	road slotting machine 路面开槽机	walk behind road slotting machine 手扶式路面开槽机
		self-propelled road slotting machine 自行式路面开槽机
	pavement crack sealing machine 路面灌缝机	towed pavement crack sealing machine, drag-type pavement crack-pouring machine 拖式路面灌缝机
		self-propelled pavement crack sealing machine 自行式路面灌缝机
	asphalt pavement heater 沥青路面加热机	self-propelled asphalt pavement heater 自行式沥青路面加热机
		towed asphalt pavement heater 拖式沥青路面加热机
		suspension asphalt pavement heater 悬挂式沥青路面加热机
	asphalt pavement heat recycling machine 沥青路面热再生机	self-propelled asphalt pavement heat recycling machine 自行式沥青路面热再生机
		towed asphalt pavement heat recycling machine 拖式沥青路面热再生机
		suspension asphalt pavement heat recycling machine 悬挂式沥青路面热再生机

（续表）

Group/组	Type/型	Product/产品
pavement maintenance machinery 路面养护机械	asphalt pavement cold recycling machine 沥青路面冷再生机	self-propelled asphalt pavement cold recycling machine 自行式沥青路面冷再生机
		towed asphalt pavement cold recycling machine 拖式沥青路面冷再生机
		suspension asphalt pavement cold recycling machine 悬挂式沥青路面冷再生机
	emulsified asphalt recycling equipment 乳化沥青再生设备	fixed emulsified asphalt recycling equipment 固定式乳化沥青再生设备
		mobile emulsified asphalt recycling equipment 移动式乳化沥青再生设备
	foam asphalt recycling equipment 泡沫沥青再生设备	fixed foam asphalt recycling equipment 固定式泡沫沥青再生设备
		mobile foam asphalt recycling equipment 移动式泡沫沥青再生设备
	chip sealing machine 碎石封层机	chip sealing machine 碎石封层机
	in-situ regenerative mixing train 就地再生搅拌列车	in-situ regenerative mixing train 就地再生搅拌列车
	pavement heater 路面加热机	pavement heater 路面加热机
	pavement heating remixer 路面加热复拌机	pavement heating remixer 路面加热复拌机
	lawn mower 割草机	lawn mower 割草机
	tree trimmer 树木修剪机	tree trimmer 树木修剪机
	road sweeper 路面清扫机	road sweeper 路面清扫机
	guardrail cleaning machine 护栏清洗机	guardrail cleaning machine 护栏清洗机

(续表)

Group/组	Type/型	Product/产品
pavement maintenance machinery 路面养护机械	construction safety indicator vehicle 施工安全指示牌车	construction safety indicator vehicle 施工安全指示牌车
	side ditch repair machine 边沟修理机	side ditch repair machine 边沟修理机
	night lighting equipment 夜间照明设备	night lighting equipment 夜间照明设备
	permeable pavement recovery machine 透水路面恢复机	permeable pavement recovery machine 透水路面恢复机
	snow removal machinery 除冰雪机械	rotor snow removal truck 转子式除雪机
		wedge plow 犁式除雪机
		spiral snowplow 螺旋式除雪机
		united snow removal truck 联合式除雪机
		snow removal truck 除雪卡车
		snowmelt spreader 融雪剂撒布机
		snowmelt sprayer 融雪液喷洒机
		spray snow remover 喷射式除冰雪机
other pavement construction and protect machinery 其他路面施工与养护机械		

7 concrete machinery 混凝土机械

Group/组	Type/型	Product/产品
concrete mixing tower 搅拌机	cone-shape reversal discharge mixer 锥形反转出料式搅拌机	gear cone-shape reversal discharge concrete mixer 齿圈锥形反转出料混凝土搅拌机
		friction cone-shape reversal discharge concrete mixer 摩擦锥形反转出料混凝土搅拌机
		diesel cone-shape reversal discharge concrete mixer 内燃机驱动锥形反转出料混凝土搅拌机
	cone-shape tipping discharge mixer 锥形倾翻出料式搅拌机	gear cone-shape tipping discharge concrete mixer 齿圈锥形倾翻出料混凝土搅拌机
		friction cone-shape tipping discharge concrete mixer 摩擦锥形倾翻出料混凝土搅拌机
		tire type full hydraulic loader 轮胎式全液压装载机
	paddle mixer 涡桨式搅拌机	paddle concrete mixer，turbo concrete mixer 涡桨式混凝土搅拌机
	planetary mixer 行星式搅拌机	planetary concrete mixer 行星式混凝土搅拌机
	single horizontal shaft mixer 单卧轴式搅拌机	single horizontal shaft mechanical feeding concrete mixer 单卧轴式机械上料混凝土搅拌机
		single horizontal shaft hydraulic feeding concrete mixer 单卧轴式液压上料混凝土搅拌机
	double horizontal shaft mixer 双卧轴式搅拌机	double horizontal shaft mechanical concrete mixer 双卧轴式机械上料混凝土搅拌机
		double horizontal shaft hydraulic concrete mixer 双卧轴式液压上料混凝土搅拌机
	continuous mixer 连续式搅拌机	continuous concrete mixer 连续式混凝土搅拌机

(续表)

Group/组	Type/型	Product/产品
concrete mixing tower, concrete batch plant, concrete mixing plant 混凝土搅拌楼	cone-shape reversal discharge mixing plant 锥形反转出料式搅拌楼	two-main-engine cone-shape reversal discharge mixing plant 双主机锥形反转出料混凝土搅拌楼
	cone-shape tipping discharge mixing plant 锥形倾翻出料式搅拌楼	two-main-engine cone-shape tipping discharge mixing plant 双主机锥形倾翻出料混凝土搅拌楼
		three-main-engine cone-shape tipping discharge mixing plant 三主机锥形倾翻出料混凝土搅拌楼
		four-main-engine cone-shape tipping discharge mixing plant 四主机锥形倾翻出料混凝土搅拌楼
	paddle mixing plant 涡桨式搅拌楼	one-main-engine paddle mixing plant 单主机涡桨式混凝土搅拌楼
		two-main-engine paddle mixing plant 双主机涡桨式混凝土搅拌楼
	planetary mixing plant 行星式搅拌楼	one-main-engine planetary concrete mixing plant 单主机行星式混凝土搅拌楼
		two-main-engine planetary concrete mixing plant 双主机行星式混凝土搅拌楼
	single horizontal shaft mixing plant 单卧轴式搅拌楼	one-main-engine single horizontal shaft concrete mixing plant 单主机单卧轴式混凝土搅拌楼
		two-main-engine single horizontal shaft concrete mixing plant 双主机单卧轴式混凝土搅拌楼
	double horizontal shaft mixing plant 双卧轴式搅拌楼	one-main-engine double horizontal shaft concrete mixing plant 单主机双卧轴式混凝土搅拌楼
		two-main-engine and double horizontal shaft concrete mixing plant 双主机双卧轴式混凝土搅拌楼
	continuous mixing plant 连续式搅拌楼	continuous concrete mixing plant 连续式混凝土搅拌楼

(续表)

Group/组	Type/型	Product/产品
concrete batch plant，concrete mixing plant，concrete mixing station 混凝土搅拌站	cone-shape reversal discharge mixing station 锥形反转出料式搅拌站	cone-shape reversal discharge concrete mixing station 锥形反转出料式混凝土搅拌站
	cone-shape tipping discharge mixing station 锥形倾翻出料式搅拌站	cone-shape tipping discharge concrete mixing station 锥形倾翻出料式混凝土搅拌站
	turboprop-type mixing station 涡桨式搅拌站	turboprop-type concrete mixing station 涡桨式混凝土搅拌站
	planetary mixing station 行星式搅拌站	planetary concrete mixing station 行星式混凝土搅拌站
	single horizontal shaft mixing station 单卧轴式搅拌站	single horizontal shaft concrete mixing station 单卧轴式混凝土搅拌站
	double horizontal shaft mixing station 双卧轴式搅拌站	double horizontal shaft concrete mixing station 双卧轴式混凝土搅拌站
	continuous mixing station 连续式搅拌站	continuous concrete mixing station 连续式混凝土搅拌站
concrete mixer truck，concrete truck mixer 混凝土搅拌运输车	self-propelled mixer truck 自行式搅拌运输车	flywheel power concrete mixer truck 飞轮取力混凝土搅拌运输车
		front-end power concrete mixer truck 前端取力混凝土搅拌运输车
		separately driven concrete mixer truck 单独驱动混凝土搅拌运输车
		front-end discharge concrete truck mixer 前端卸料混凝土搅拌运输车
		concrete truck mixer with belt conveyer 带皮带输送机混凝土搅拌运输车

（续表）

Group/组	Type/型	Product/产品
concrete mixer truck，concrete truck mixer 混凝土搅拌运输车	self-propelled mixer truck 自行式搅拌运输车	concrete mixer truck with feeding device 带上料装置混凝土搅拌运输车
		concrete mixer truck with boom concrete pump 带臂架混凝土泵混凝土搅拌运输车
		concrete mixer truck with tipping mechanism 带倾翻机构混凝土搅拌运输车
	towed mixer truck 拖式搅拌运输车	concrete mixer truck，concrete truck mixer 混凝土搅拌运输车
concrete pump 混凝土泵	stationary pump 固定式泵	stationary concrete pump 固定式混凝土泵
	towed pump 拖式泵	towed concrete pump，trailer concrete pump 拖式混凝土泵
	truck-mounted pump 车载式泵	truck-mounted concrete pump 车载式混凝土泵
concrete placing boom 混凝土布料杆	foldable placing boom，foldable distributor boom 卷折式布料杆	foldable concrete placing boom 卷折式混凝土布料杆
	Z-shaped folding placing boom，Z-shaped folding distributor boom “Z”形折叠式布料杆	Z-shaped folding concrete placing boom “Z”形折叠式混凝土布料杆
	telescopic placing boom，telescopic distributor boom 伸缩式布料杆	telescopic concrete placing boom 伸缩式混凝土布料杆
	combined placing boom，combined distributor boom 组合式布料杆	foldable Z-shaped folding composite concrete placing boom 卷折“Z”形折叠组合式混凝土布料杆
		Z-shaped folding and telescopic composite concrete placing boom “Z”形折叠伸缩组合式混凝土布料杆
		folding and telescopic composite concrete placing boom 卷折伸缩组合式混凝土布料杆

(续表)

Group/组	Type/型	Product/产品
jib concrete pump truck, truck mounted concrete pump with placing boom 臂架式混凝土泵车	comprehensive pump truck 整体式泵车	comprehensive jib concrete pump truck 整体式臂架式混凝土泵车
jib concrete pump truck, truck mounted concrete pump with placing boom 臂架式混凝土泵车	semi-mounted pump truck 半挂式泵车	semi-mounted jib concrete pump truck 半挂式臂架式混凝土泵车
jib concrete pump truck, truck mounted concrete pump with placing boom 臂架式混凝土泵车	full-mounted pump truck 全挂式泵车	full-mounted jib concrete pump truck 全挂式臂架式混凝土泵车
screw concrete spraying machine, spiral concrete sprayer 混凝土喷射机	cylinder sprayer 缸罐式喷射机	cylinder concrete sprayer 缸罐式混凝土喷射机
screw concrete spraying machine, spiral concrete sprayer 混凝土喷射机	screw spraying machine 螺旋式喷射机	screw concrete spraying machine, spiral concrete sprayer 螺旋式混凝土喷射机
screw concrete spraying machine, spiral concrete sprayer 混凝土喷射机	rotor sprayer 转子式喷射机	rotor concrete sprayer, rotor concrete spraying machine 转子式混凝土喷射机
concrete jet jack manipulator 混凝土喷射机械手	concrete jet jack manipulator 混凝土喷射机械手	concrete jet jack manipulator 混凝土喷射机械手
concrete jet jack plane 混凝土喷射台车	concrete jet jack plane 混凝土喷射台车	concrete jet jack plane 混凝土喷射台车
concrete costing machine 混凝土浇注机	rail mounted casting machine 轨道式浇注机	rail mounted concrete casting machine 轨道式混凝土浇注机
concrete costing machine 混凝土浇注机	tire casting machine 轮胎式浇注机	tire mounted concrete casting machine 轮胎式混凝土浇注机
concrete costing machine 混凝土浇注机	fixed casting machine 固定式浇注机	fixed concrete casting machine 固定式混凝土浇注机
concrete vibrator, vibrator for compacting concrete mix 混凝土振动器	internal vibrator 内部振动式振动器	electric planetary soft-shaft insert concrete vibrator 电动软轴行星插入式混凝土振动器
concrete vibrator, vibrator for compacting concrete mix 混凝土振动器	internal vibrator 内部振动式振动器	electric eccentric soft-shaft insert concrete vibrator 电动软轴偏心插入式混凝土振动器

(续表)

Group/组	Type/型	Product/产品
concrete vibrator, vibrator for compacting concrete mix 混凝土振动器	internal vibrator 内部振动式振动器	diesel planetary soft-shaft insert concrete vibrator 内燃软轴行星插入式混凝土振动器
		inbuilt motor insert concrete vibrator 电机内装插入式混凝土振动器
	external vibrator 外部振动式振动器	flat concrete vibrator 平板式混凝土振动器
		attached concrete vibrator 附着式混凝土振动器
		unidirectional vibrating attached concrete vibrator 单向振动附着式混凝土振动器
concrete vibrating table 混凝土振动台	concrete vibrating table 混凝土振动台	concrete vibrating table 混凝土振动台
bulk-cement pneumatic delivery tanker 气卸散装水泥运输车	bulk-cement pneumatic delivery tanker 气卸散装水泥运输车	bulk-cement pneumatic delivery tanker 气卸散装水泥运输车
concrete cleaning recycling station 混凝土清洗回收站	concrete cleaning recycling station 混凝土清洗回收站	concrete cleaning recycling station 混凝土清洗回收站
concrete batching plant 混凝土配料站	concrete batching plant 混凝土配料站	concrete batching plant 混凝土配料站
other concrete machinery 其他混凝土机械		

8 boring machinery 掘进机械

Group/组	Type/型	Product/产品
tunnel boring machine(TBM) 全断面隧道掘进机	shield machine, tunnel boring machine 盾构机	earth pressure balanced shield machine 土压平衡式盾构机
		sludge balanced shield machine 泥水平衡式盾构机

(续表)

Group/组	Type/型	Product/产品
tunnel boring machine(TBM) 全断面隧道掘进机	shield machine, tunnel boring machine 盾构机	slurry shield machine 泥浆式盾构机
		sludge shield machine 泥水式盾构机
		special-shaped shield machine 异型盾构机
	hard rock tunnel boring machine 硬岩掘进机	hard rock tunnel boring machine 硬岩掘进机
	combined boring machine 组合式掘进机	combined boring machine 组合式掘进机
trenchless equipments 非开挖设备	horizontal directional drill 水平定向钻	horizontal directional drill 水平定向钻
	pipe jacking machine 顶管机	earth pressure balanced pipe jacking machine 土压平衡式顶管机
		sludge balanced pipe jacking machine 泥水平衡式顶管机
		slurry conveying pipe jacking machine 泥水输送式顶管机
tunnel boring machine 巷道掘进机	cantilever rock roadheader 悬臂式岩巷掘进机	cantilever rock roadheader 悬臂式岩巷掘进机
other boring machinery 其他掘进机械		

9 pile driving machine 桩工机械

Group/组	Type/型	Product/产品名称
diesel piling hammer 柴油打桩锤	tubular piling hammer 筒式打桩锤	cold water tubular diesel piling hammer 水冷筒式柴油打桩锤
		cold wind tubular diesel piling hammer 风冷筒式柴油打桩锤

（续表）

Group/组	Type/型	Product/产品名称
diesel piling hammer 柴油打桩锤	slide bar piling hammer 导杆式打桩锤	slide bar diesel piling hammer 导杆式柴油打桩锤
hydraulic hammer, hydraulically powered impact hammer 液压锤	hydraulic hammer, hydraulically powered impact hammer 液压锤	hydraulic pile hammer 液压打桩锤
vibration hammer, vibrator for piling equipment, vibratory pile hammer 振动桩锤	mechanical pile hammer 机械式桩锤	common vibratory pile hammer 普通振动桩锤
		variable torque vibratory pile hammer 变矩振动桩锤
		variable frequency vibratory pile hammer 变频振动桩锤
		variable torque and frequency vibratory pile hammer 变矩变频振动桩锤
	hydraulic motor pile hammer 液压马达式桩锤	hydraulic motor vibratory pile hammer 液压马达式振动桩锤
	hydraulic pile hammer 液压式桩锤	hydraulic vibrator for piling equipment, hydraulic vibratory pile hammer, vibrating pile driver (hydraulic type) 液压振动锤
piling frame 桩架	pipe piling frame 走管式桩架	pipe piling frame of diesel hammer 走管式柴油打桩架
	orbit piling frame 轨道式桩架	orbit piling frame of diesel hammer 轨道式柴油锤打桩架
	track-laying piling frame 履带式桩架	track-laying piling frame of diesel hammer in tri-point support 履带三支点式柴油锤打桩架
	walking type piling frame 步履式桩架	walking type piling frame 步履式桩架
	suspension pile frame 悬挂式桩架	crawler suspended diesel hammer pile frame 履带悬挂式柴油锤桩架

(续表)

Group/组	Type/型	Product/产品名称
pile driver 压桩机	mechanical pile driver，mechanical pile pressing machine 机械式压桩机	mechanical pile driver，mechanical pile pressing machine 机械式压桩机
	hydraulic pile driver，hydraulic press pile driver 液压式压桩机	hydraulic pile driver，hydraulic press pile driver 液压式压桩机
boring holes machine，drilling machine 成孔机	auger drilling machine 螺旋式成孔机	long auger drilling machine 长螺旋钻孔机
		extrusion long auger drilling machine 挤压式长螺旋钻孔机
		telescopic long auger drilling machine 套管式长螺旋钻孔机
		short auger drilling machine 短螺旋钻孔机
	dive boring holes machine 潜水式成孔机	dive boring holes machine 潜水钻孔机
	positive and negative rotary drilling machine 正反回转式成孔机	rotary drilling machine 转盘式钻孔机
		top-drive drilling machine 动力头式钻孔机
	punching-grab boring holes machine 冲抓式成孔机	punching-grab boring holes machine 冲抓成孔机
	all-casing drilling machine 全套管式成孔机	all-casing drilling machine 全套管钻孔机
	bolt drilling machine 锚杆式成孔机	bolt drilling machine 锚杆钻孔机
	walking drilling machine. 步履式成孔机	walking rotary drilling machine 步履式旋挖钻孔机
	crawler drilling machine 履带式成孔机	crawler rotary drilling machine 履带式旋挖钻孔机

（续表）

Group/组	Type/型	Product/产品名称
drilling machine boring holes machine 成孔机	vehicle-mounted drilling machine 车载式成孔机	vehicle-mounted rotary drilling machine 车载式旋挖钻孔机
	multi-shaft drilling machine 多轴式成孔机	multi-shaft drilling machine 多轴钻孔机
grooving machine for underground continuous wall 地下连续墙成槽机	wire rope grooving machine 钢丝绳式成槽机	mechanical diaphragm wall grab 机械式连续墙抓斗
	guide rod grooving machine 导杆式成槽机	hydraulic diaphragm wall grab 液压式连续墙抓斗
	semi-guide grooving machine 半导杆式成槽机	hydraulic diaphragm wall grab 液压式连续墙抓斗
	milling grooving machine 铣削式成槽机	double wheels trench cutter 双轮铣成槽机
	agitating grooving machine 搅拌式成槽机	double wheels mixer 双轮搅拌机
	diving grooving machine 潜水式成槽机	diving vertical multi-axis grooving machine 潜水式垂直多轴成槽机
drop hammer pile driver 落锤打桩机	mechanical pile driver 机械式打桩机	mechanical drop hammer pile driver 机械式落锤打桩机
	frank pile driver 法兰克式打桩机	frank pile driver 法兰克式打桩机
consolidation machinery for soft ground 软地基加固机械	vibroflotation consolidation machinery 振冲式加固机械	hydraulic vibrator 水冲式振冲器
		dry vibrator 干式振冲器
	plug-in consolidation machinery 插板式加固机械	plug pile machine 插板桩机
	dynamic compaction consolidation machinery 强夯式加固机械	dynamic compaction machinery 强夯机

(续表)

Group/组	Type/型	Product/产品名称
consolidation machinery for soft ground 软地基加固机械	vibration consolidation machinery 振动式加固机械	sand pile machine 砂桩机
	rotary spraying consolidation machinery 旋喷式加固机械	rotary spraying consolidation machinery for soft ground 旋喷式软地基加固机
	grouting deep mixing consolidation machinery 注浆式深层搅拌式加固机械	single shaft grouting deep mixer 单轴注浆式深层搅拌机
		multi-shaft grouting deep mixer 多轴注浆式深层搅拌机
	powder jet deep mixing consolidation machinery 粉体喷射式深层搅拌式加固机械	single shaft powder jet deep mixing consolidation machinery 单轴粉体喷射式深层搅拌机
		multi-shaft powder jet deep mixing consolidation machinery 多轴粉体喷射式深层搅拌机
soil sampler 取土器	thick-wall soil sampler 厚壁取土器	thick-wall soil sampler 厚壁取土器
	open thin-walled soil sampler 敞口薄壁取土器	open thin-walled soil sampler 敞口薄壁取土器
	free piston thin-walled soil sampler 自由活塞薄壁取土器	free piston thin-walled soil sampler 自由活塞薄壁取土器
	fixed piston thin-walled soil sampler 固定活塞薄壁取土器	fixed piston thin-walled soil sampler 固定活塞薄壁取土器
	hydraulically fixed piston soil collector 水压固定薄壁取土器	hydraulically fixed piston soil collector 水压固定薄壁取土器
	sampler with thin-walled extension of shoe 束节式取土器	sampler with thin-walled extension of shoe 束节式取土器
	loess sampler 黄土取土器	loess sampler 黄土取土器

(续表)

Group/组	Type/型	Product/产品名称
soil sampler 取土器	triple tube rotary corer 三重管回转式取土器	triple tube swivel type rotary corer 三重管单动回转取土器
		triple tube rigid type rotary corer 三重管双动回转取土器
	sand sampler 取砂器	undisturbed sand sampler 原状取沙器
other pile driving machine 其他桩工机械		

10 municipal engineering and sanitation machinery 市政与环卫机械

Group/组	Type/型	Product/产品
sanitation machinery 环卫机械	sweep road car 扫路车(机)	sweep road car 扫路车
		sweeper 扫路机
	vacuum sweeper 吸尘车	vacuum sweeper 吸尘车
	cleaning sweeper truck 洗扫车	cleaning sweeper truck 洗扫车
	cleaning vehicle 清洗车	cleaning vehicle 清洗车
		guardrail cleanout vehicle 护栏清洗车
		wall sweeper 洗墙车
	motor sprinkler, sprinkler, sprinkler truck, street sprinkler, water dispenser, watering cart 洒水车	motor sprinkler, sprinkler, sprinkler truck, street sprinkler, water dispenser, watering cart 洒水车
		cleaning sprinkler 清洗洒水车
		tree sprinkling tanker 绿化喷洒车

（续表）

Group/组	Type/型	Product/产品
sanitation machinery 环卫机械	fecal suction truck 吸粪车	fecal suction truck 吸粪车
	toilet car 厕所车	toilet car 厕所车
	waste collecting vehicle 垃圾车	compression refuse collector 压缩式垃圾车
		garbage dump truck 自卸式垃圾车
		garbage collector 垃圾收集车
		dump garbage collector 自卸式垃圾收集车
		tricycle garbage collector 三轮垃圾收集车
		self-loading garbage truck 自装卸式垃圾车
		swept body refuse collector 摆臂式垃圾车
		detachable container garbage collector 车厢可卸式垃圾车
		sorting garbage truck 分类垃圾车
		compressed sorting garbage truck 压缩式分类垃圾车
		garbage carrying vehicle 垃圾转运车
		barreled garbage truck 桶装垃圾运输车
		kitchen garbage vehicle 餐厨垃圾车
		medical garbage truck 医疗垃圾车
	refuse handling installation 垃圾处理设备	rubbish compressor 垃圾压缩机
		crawler type waste bulldozer 履带式垃圾推土机
		crawler type waste excavator 履带式垃圾挖掘机

（续表）

Group/组	Type/型	Product/产品
sanitation machinery 环卫机械	refuse handling installation 垃圾处理设备	landfill leachate treatment vehicle 垃圾渗滤液处理车
		waste transfer station equipment 垃圾中转站设备
		waste screening machine 垃圾分拣机
		waste incinerator 垃圾焚烧炉
		waste crusher 垃圾破碎机
		waste composting equipment 垃圾堆肥设备
		waste incineration equipment 垃圾填埋设备
municipal machinery 市政机械	pipe dredging machine 管道疏通机械	sewage truck 吸污车
		cleaning suction truck 清洗吸污车
urban construction machinery 市政机械	pipe dredging machine 管道疏通机械	sewer integrated maintenance vehicle 下水道综合养护车
		sewer dredge vehicle 下水道疏通车
		sewer dredging and cleaning vehicle 下水道疏通清洗车
		excavating machine 掏挖车
		sewer repair and inspection equipment 下水道检查修补设备
		sludge truck 污泥运输车
	electric pole embedding machinery 电杆埋架机械	electric pole embedding machinery 电杆埋架机械
	pipelayer 管道铺设机械	pipelayer 铺管机

(续表)

Group/组	Type/型	Product/产品
parking and car washing equipment 停车洗车设备	vertical circulating parking equipment 垂直循环式停车设备	vertical circulation lower access parking equipment 垂直循环式下部出入式停车设备
		vertical circulation central access parking equipment 垂直循环式中部出入式停车设备
		vertical circulation upper access parking equipment 垂直循环式上部出入式停车设备
	multilaye circulating parking equipment 多层循环式停车设备	multilaye circular circulating parking equipment 多层圆形循环式停车设备
		multilaye rectangular circulating parking equipment 多层矩形循环式停车设备
	horizontal circulating parking equipment 水平循环式停车设备	horizontal circular circulating parking equipment 水平圆形循环式停车设备
		horizontal rectangular circulating parking equipment 水平矩形循环式停车设备
	lift parking equipment 升降机式停车设备	lift vertical parking equipment 升降机纵置式停车设备
		lift horizontal parking equipment 升降机横置式停车设备
		lift circular parking equipment 升降机圆置式停车设备
	lift mobile parking equipment 升降移动式停车设备	lift mobile vertical parking equipment 升降移动纵置式停车设备
		lift mobile horizontal parking equipment 升降移动横置式停车设备
	planar reciprocating parking equipment 平面往复式停车设备	planar reciprocating handling parking equipment 平面往复搬运式停车设备
		planar reciprocating handling and receiving parking equipment 平面往复搬运收容式停车设备

（续表）

Group/组	Type/型	Product/产品
parking and car washing equipment 停车洗车设备	two-storey parking equipment 两层式停车设备	two-storey lifting parking equipment 两层升降式停车设备
		two-storey lifting and transverse parking equipment 两层升降横移式停车设备
	multilayer parking equipment 多层式停车设备	multilayer lifting parking equipment 多层升降式停车设备
		multilayer lifting and transverse parking equipment 多层升降横移式停车设备
	automobile rotary disc parking equipment 汽车用回转盘停车设备	automobile rotary turntable 旋转式汽车用回转盘
		automobile rotary mobile turntable 旋转移动式汽车用回转盘
	car lift parking equipment 汽车用升降机停车设备	elevator for automobile 升降式汽车用升降机
		lift-rotary elevator for automobile 升降回转式汽车用升降机
		lift-sliding elevator for automobile 升降横移式汽车用升降机
	revolving platform parking equipment 旋转平台停车设备	revolving platform 旋转平台
	car wash machinery 洗车场机械设备	car wash machinery 洗车场机械设备
gardening machinery, horticultural machine 园林机械	digging machine for planting trees 植树挖穴机	self-propelled digging machine for planting trees 自行式植树挖穴机
		walk behind digging machine for planting trees 手扶式植树挖穴机
	tree transplanter 树木移植机	self-propelled tree transplanter 自行式树木移植机
		towed tree transplanter 牵引式树木移植机
		suspension tree transplanter 悬挂式树木移植机

(续表)

Group/组	Type/型	Product/产品
gardening machinery, horticultural machine 园林机械	tree carrying vehicle 运树机	multi-bucket towed tree carrying vehicle 多斗拖挂式运树机
	multi-purpose spraying vehicle 绿化喷洒多用车	hydraulic spray type multi-purpose spraying vehicle 液力喷雾式绿化喷洒多用车
gardening machinery, horticultural machine 园林机械	lawn mower 剪草机	hand-propelled rotary lawn mower 手推式旋刀剪草机
		towed cylind lawn mower 拖挂式滚刀剪草机
		riding cylind lawn mower 乘座式滚刀剪草机
		self-propelled cylind lawn mower 自行式滚刀剪草机
		hand-propelled cylind lawn mower 手推式滚刀剪草机
		self-propelled reciprocating lawn mower 自行式往复剪草机
		hand-propelled reciprocating lawn mower 手推式往复剪草机
		flail type lawn mower 甩刀式剪草机
		air cushion type lawn mower 气垫式剪草机
recreational equipment 娱乐设备	car entertainment equipment 车式娱乐设备	race car 小赛车
		bumper car 碰碰车
		ferris wheel 观览车
		battery car 电瓶车
		sightseeing vehicle 观光车
	water entertainment equipment 水上娱乐设备	battery boat 电瓶船
		pedal boat 脚踏船

（续表）

Group/组	Type/型	Product/产品
recreational equipment 娱乐设备	water entertainment equipment 水上娱乐设备	bumper boat 碰碰船
		lotic brave boat 激流勇进船
		water yacht 水上游艇
	ground entertainment equipment 地面娱乐设备	amusement machine 游艺机
		trampoline 蹦床
		merry go round 转马
		go by like the wind 风驰电掣
	aerial entertainment equipment 腾空娱乐设备	rotary astro fighter 旋转自控飞机
		moon shuttle 登月火箭
		air swivel chair 空中转椅
		space travel 宇宙旅行
	other entertainment equipment 其他娱乐设备	other entertainment equipment 其他娱乐设备
other municipal and sanitation machinery 其他市政与环卫机械		

11　concrete products machinery 混凝土制品机械

Group/组	Type/型	Product/产品
concrete block making machine, concrete block forming machine 混凝土砌块成型机	mobile 移动式	mobile hydraulic demoulding concrete block making machine 移动式液压脱模混凝土砌块成型机

(续表)

Group/组	Type/型	Product/产品
concrete block making machine, concrete block forming machine 混凝土砌块成型机	mobile 移动式	mobile mechanical demoulding concrete block making machine 移动式机械脱模混凝土砌块成型机
		mobile artificial demoulding concrete block making machine 移动式人工脱模混凝土砌块成型机
	fixed 固定式	fixed mode vibration hydraulic demoulding concrete block making machine 固定式模振液压脱模混凝土砌块成型机
		fixed mode vibration mechanical demoulding concrete block making machine 固定式模振机械脱模混凝土砌块成型机
		fixed mode vibration artificial demoulding concrete block making machine 固定式模振人工脱模混凝土砌块成型机
		fixed bench vibration hydraulic demoulding concrete block making machine 固定式台振液压脱模混凝土砌块成型机
concrete block making machine 混凝土砌块成型机	fixed 固定式	fixed bench vibration mechanical demoulding concrete block making machine 固定式台振机械脱模混凝土砌块成型机
		fixed bench vibration manual demoulding concrete block making machine 固定式台振人工脱模混凝土砌块成型机
	laminated 叠层式	laminated concrete block making machine 叠层式混凝土砌块成型机
	layered-filling 分层布料式	layered-filling concrete block making machine 分层布料式混凝土砌块成型机

(续表)

Group/组	Type/型	Product/产品
complete equipments for producing concrete block 混凝土砌块生产成套设备	full-automatic 全自动	full-automatic bench vibration complete equipments for producing concrete block 全自动台振混凝土砌块生产线
		full-automatic mode vibration complete equipments for producing concrete block 全自动模振混凝土砌块生产线
	semi-automatic 半自动	semi-automatic bench vibration complete equipments for producing concrete block 半自动台振混凝土砌块生产线
		semi-automatic mode vibration complete equipments for producing concrete block 半自动模振混凝土砌块生产线
	simple type 简易式	simple bench vibration complete equipments for producing concrete block 简易台振混凝土砌块生产线
		simple bench vibration complete equipments for producing concrete block 简易模振混凝土砌块生产线
autoclaved aerated concrete equipment 加气混凝土砌块成套设备	autoclaved aerated concrete equipment 加气混凝土砌块设备	autoclaved aerated concrete block production line 加气混凝土砌块生产线
foamed concrete equipment 泡沫混凝土砌块成套设备	foamed concrete equipment 泡沫混凝土砌块设备	foam concrete block forming machine 泡沫混凝土砌块生产线
concrete hollow slab forming machine 混凝土空心板成型机	screw 挤压式	external vibration type screw extruder for producing single concrete hollow slab 外振式单块混凝土空心板挤压成型机
		external vibration type screw extruder for producing double concrete hollow slab 外振式双块混凝土空心板挤压成型机

(续表)

Group/组	Type/型	Product/产品
concrete hollow slab forming machine 混凝土空心板成型机	screw 挤压式	internal vibration type screw extruder for producing single concrete hollow slab 内振式单块混凝土空心板挤压成型机
		internal vibration type screw extruder for producing double concrete hollow slab 内振式双块混凝土空心板挤压成型机
	pushing 推压式	external vibration type pushing extruder for producing single concrete hollow slab 外振式单块混凝土空心板推压成型机
		external vibration type pushing extruder for producing double concrete hollow slab 外振式双块混凝土空心板推压成型机
		internal vibration type pushing extruder for producing single concrete hollow slab 内振式单块混凝土空心板推压成型机
		internal vibration type pushing extruder for producing double concrete hollow slab 内振式双块混凝土空心板推压成型机
	mould dragger 拉模式	self-propelled external vibration mould dragger for producing concrete hollow slab 自行式外振混凝土空心板拉模成型机
		towed external vibration mould dragger for producing concrete hollow slab 牵引式外振混凝土空心板拉模成型机
		self-propelled internal vibration mould dragger for producing concrete hollow slab 自行式内振混凝土空心板拉模成型机
		towed internal vibration mould dragger for producing concrete hollow slab 牵引式内振混凝土空心板拉模成型机

(续表)

Group/组	Type/型	Product/产品
concrete component forming machine 混凝土构件成型机	vibrating table forming machine 振动台式成型机	electric vibration table concrete component form machine 电动振动台式混凝土构件成型机
		pneumatic vibration table concrete component form machine 气动振动台式混凝土构件成型机
		non-bench vibration table concrete component form machine 无台架振动台式混凝土构件成型机
		horizontal directional vibration table concrete component form machine 水平定向振动台式混凝土构件成型机
		shock vibration table concrete component form machine 冲击振动台式混凝土构件成型机
		roller pulse vibration table concrete component form machine 滚轮脉冲振动台式混凝土构件成型机
		segmented combined vibration table concrete component form machine 分段组合振动台式混凝土构件成型机
	disk-turning compression molding machine 盘转压制式成型机	concrete component disk-turning compression molding machine 混凝土构件盘转压制成型机
	lever compression moulding machine 杠杆压制式成型机	concrete component lever compression molding machine 混凝土构件杠杆压制成型机
	long-line pedestal 长线台座式	long-line pedestal concrete component production equipment 长线台座式混凝土构件生产成套设备
	flat-mode linkage 平模联动式	flat mold linkage concrete component production equipment 平模联动式混凝土构件生产成套设备
	unit linkage 机组联动式	unit linkage concrete component production equipment 机组联动式混凝土构件生产成套设备

(续表)

Group/组	Type/型	Product/产品
concrete pipe forming machine 混凝土管成型机	centrifugal type 离心式	roller centrifugal concrete pipe forming machine 滚轮离心式混凝土管成型机
		lathe centrifugal concrete pipe forming machine 车床离心式混凝土管成型机
	extrusion type 挤压式	hanging roller extrusion concrete pipe forming machine 悬辊式挤压混凝土管成型机
		vertical extrusion concrete pipe forming machine 立式挤压混凝土管成型机
		vertical vibrating extruded concrete pipe forming machine 立式振动挤压混凝土管成型机
cement tile forming machine 水泥瓦成型机	cement tile forming machine 水泥瓦成型机	cement tile forming machine 水泥瓦成型机
wallboard forming machine 墙板成型设备	wallboard forming machine 墙板成型机	wallboard forming machine 墙板成型机
concrete component repair machine 混凝土构件修整机	vacuum suction device 真空吸水装置	concrete vacuum suction device 混凝土真空吸水装置
	cutter, cutting machine, slicer 切割机	walk behind concrete cutter 手扶式混凝土切割机
		self-propelled concrete cutter 自行式混凝土切割机
	surface polisher 表面抹光机	walk behind concrete surface polisher 手扶式混凝土表面抹光机
		self-propelled concrete surface polisher 自行式混凝土表面抹光机
	grinding machine 磨口机	concrete pipe grinding machine 混凝土管件磨口机
formwork and accessories machinery 模板及配件机械	steel formwork mill 钢模板轧机	steel formwork continuous mill 钢模版连轧机
		steel formwork ridge mill 钢模板凸棱轧机

(续表)

Group/组	Type/型	Product/产品
formwork and accessories machinery 模板及配件机械	steel form cleaning machine 钢模板清理机	steel form cleaning machine 钢模板清理机
	steel formwork calibration machine 钢模板校形机	steel formwork multi-function calibration machine 钢模板多功能校形机
	steel formwork accessories 钢模板配件	U-shaped card forming machine for steel formwork 钢模板U形卡成型机
		straightening machine for steel formwork steel pipe 钢模板钢管校直机
other concrete products machinery 其他混凝土制品机械		

12 aerial work machine, aerial working machinery 高空作业机械

Group/组	Type/型	Product/产品
aerial work vehicle, aerial working truck, overhead working truck, vehicle for high lift work, vehicle-mounted mobile elevating work platform 高空作业车	ordinary aerial work vehicle 普通型高空作业车	telescopic aerial working vehicle 伸臂式高空作业车
		folding jib aerial working vehicle 折叠臂式高空作业车
		vertical lifts aerial work vehicle 垂直升降式高空作业车
		hybrid aerial work vehicle 混合式高空作业车
	pruning vehicle, trim car 高树剪枝车	pruning vehicle, trim car 高树剪枝车
		trailer-mounted pruning vehicle, trailer-mounted trim car 拖式高空剪枝车
	aerial insulated vehicle 高空绝缘车	aerial insulated bucket boom truck 高空绝缘斗臂车
		trailer-mounted aerial insulated vehicle 拖式高空绝缘车

（续表）

Group/组	Type/型	Product/产品
aerial work vehicle，aerial working truck，overhead working truck，vehicle for high lift work，vehicle-mounted mobile elevating work platform 高空作业车	bridge inspection plant 桥梁检修设备	bridge inspection truck 桥梁检修车
		trailer-mounted bridge inspection platform 拖式桥梁检修平台
	aerial dolly 高空摄影车	aerial dolly 高空摄影车
	aviation ground support vehicle 航空地面支持车	aviation ground support lift vehicle 航空地面支持用升降车
	aircraft deicing vehicle 飞机除冰防冰车	aircraft deicing vehicle 飞机除冰防冰车
	emergency tender，emergency truck 消防救援车	aerial emergency tender，aerial emergency truck 高空消防救援车
aerial work platform 高空作业平台	scissors aerial work platform 剪叉式高空作业平台	stationary scissors aerial work platform 固定剪叉式高空作业平台
		mobile scissors aerial work platform 移动剪叉式高空作业平台
		self-propelled scissors aerial work platform 自行剪叉式高空作业平台
	arm aerial work platform 臂架式高空作业平台	stationary arm aerial work platform 固定臂架式高空作业平台
		mobile arm aerial work platform 移动臂架式高空作业平台
		self-propelled arm aerial work platform 自行臂架式高空作业平台
	sleeving cylinder aerial work platform，telescopic cylindrical aerial work platform 套筒油缸式高空作业平台	stationary sleeving cylinder aerial work platform，stationary telescopic cylindrical aerial work platform 固定套筒油缸式高空作业平台
		mobile sleeving cylinder aerial work platform，mobile telescopic cylindrical aerial work platform 移动套筒油缸式高空作业平台

（续表）

Group/组	Type/型	Product/产品
aerial work platform 高空作业平台	sleeving cylinder aerial work platform, telescopic cylindrical aerial work platform 套筒油缸式高空作业平台	self- propelled sleeving cylinder aerial work platform, self-propelled telescopic cylindrical aerial work platform 自行套筒油缸式高空作业平台
	mast aerial work platform 桅柱式高空作业平台	stationary mast aerial work platform 固定桅柱式高空作业平台
		mobile mast aerial work platform 移动桅柱式高空作业平台
		self-propelled mast aerial work platform 自行桅柱式高空作业平台
	guide rail aerial work platform, truss aerial work platform 导架式高空作业平台	stationary truss aerial work platform, stationary guide rail aerial work platform 固定导架式高空作业平台
		mobile truss aerial work platform, mobile guide rail aerial work platform 移动导架式高空作业平台
		self-propelled truss aerial work platform/self-propelled guide rail aerial work platform 自行导架式高空作业平台
other aerial work platform 其他高空作业机械		

13 decoration machinery, finishing machinery 装修机械

Group/组	Type/型	Product/产品
mortar preparation and painting machinery 砂浆制备及喷涂机械	sand screener, sand sieving machine, sand sifter 筛砂机	electric sieve machine 电动式筛砂机
	haxis mortar mixer 砂浆搅拌机	horizontal axis cement mortar mixer 卧轴式灰浆搅拌机

(续表)

Group/组	Type/型	Product/产品
mortar preparation and painting machinery 砂浆制备及喷涂机械	haxis mortar mixer 砂浆搅拌机	vertical axis cement mortar mixer 立轴式灰浆搅拌机
		rotary cylinder cement mortar mixer 筒转式灰浆搅拌机
	mortar trasmission pump 泵浆输送泵	piston single-cylinder cement mortar pump 柱塞式单缸灰浆泵
		piston twin-cylinder cement mortar pump 柱塞式双缸灰浆泵
		diaphragm cement mortar pump 隔膜式灰浆泵
		pneumatic cement mortar pump 气动式灰浆泵
		squeeze cement mortar pump 挤压式灰浆泵
		screw cement mortar pump 螺杆式灰浆泵
	cement mortar joint machine 砂浆联合机	cement mortar joint machine 灰浆联合机
	ash dryer 淋灰机	ash dryer 淋灰机
	hemp knife ash mixer 麻刀灰拌和机	hemp knife ash mixer 麻刀灰拌和机
spraying machine, paint sprayer 涂料喷刷机械	pulp pump, gunite pump 喷浆泵	pulp pump, gunite pump 喷浆泵
	airless paint sprayer 无气喷涂机	pneumatic airless paint sprayer 气动式无气喷涂机
		electric airless paint sprayer 电动式无气喷涂机
		internal combustion airless paint sprayer 内燃式无气喷涂机
		high pressure airless paint sprayer 高压无气喷涂机

(续表)

Group/组	Type/型	Product/产品
spraying machine，paint sprayer 涂料喷刷机械	air sprayer 有气喷涂机	exhaust type air sprayer 抽气式有气喷涂机
		down-type air sprayer 自落式有气喷涂机
	plastic spraying machine 喷塑机	plastic spraying machine 喷塑机
	plaster spraying machine 石膏喷涂机	plaster spraying machine 石膏喷涂机
painting machine and spraying machine 油漆制备及喷涂机械	paint spraying machine 油漆喷涂机	paint spraying machine 油漆喷涂机
	paint mixer，painting mixer 油漆搅拌机	paint mixer，painting mixer 油漆搅拌机
floor finishing machine 地面修整机械	surface polishing 地面抹光机	surface polishing 地面抹光机
	floor grinder，floor polish 地板磨光机	floor grinder，floor polish 地板磨光机
	skirt line grinder 踢脚线磨光机	skirt line grinder 踢脚线磨光机
	the ground terrazzo machine 地面水磨石机	single-plate terrazzo machine 单盘水磨石机
		double-plate terrazzo machine 双盘水磨石机
		diamond ground terrazzo machine 金刚石地面水磨石机
	floor-planning，plank floor planer 地板刨平机	floor-planning，plank floor planer 地板刨平机
	waxing machine 打蜡机	waxing machine 打蜡机
	floor cleaner 地面清除机	floor cleaner 地面清除机
	floor tile cutter，floor tile cutting machine 地板砖切割机	floor tile cutter，floor tile cutting machine 地板砖切割机

(续表)

Group/组	Type/型	Product/产品
roof decoration machinery 屋面装修机械	asphalt coating machine 涂沥青机	asphalt coating machine for roofing 屋面涂沥青机
	felting machine 铺毡机	felting machine for roofing 屋面铺毡机
high place work gondola 高处作业吊篮	manual high-altitude work basket 手动式高处作业吊篮	manual high-altitude work basket 手动高处作业吊篮
	pneumatic high-altitude work basket 气动式高处作业吊篮	pneumatic high-altitude work basket 气动高处作业吊篮
	electric climbing high-altitude work basket 电动式高处作业吊篮	electric climbing rope type high-altitude work basket 电动爬绳式高处作业吊篮
		electrically operated hoist type high-altitude work basket 电动卷扬式高处作业吊篮
permanently installed suspended access equipment, window-cleaning machine 擦窗机	wheel-mounted window cleaner 轮毂式擦窗机	wheel-mounted telescope jib variable amplitude window cleaner 轮毂式伸缩变幅擦窗机
		wheel-mounted wheelbarrow variable amplitude window cleaner 轮毂式小车变幅擦窗机
		wheel-mounted arm variable amplitude window cleaner 轮毂式动臂变幅擦窗机
	roof track window cleaner 屋面轨道式擦窗机	roof track telescope jib variable amplitude window cleaner 屋面轨道式伸缩臂变幅擦窗机
		roof track wheelbarrow variable amplitude window cleaner 屋面轨道式小车变幅擦窗机
		roof track arm variable amplitude window cleaner 屋面轨道式动臂变幅擦窗机
	suspended track window cleaner 悬挂轨道式擦窗机	suspended track window cleaner 悬挂轨道式擦窗机

（续表）

Group/组	Type/型	Product/产品
permanently installed suspended access equipment，window-cleaning machine 擦窗机	insertion-type window clean machine 插杆式擦窗机	insertion-type window clean machine 插杆式擦窗机
	slide window clean machine 滑梯式擦窗机	slide window clean machine 滑梯式擦窗机
building decoration machine 建筑装修机具	nailing machine 射钉机	nailing machine 射钉机
	scaling scraper 铲刮机	electric scaling scraper 电动铲刮机
	notching machine 开槽机	concrete notching machine 混凝土开槽机
	plain stone cutter，stone cutting machine 石材切割机	plain stone cutter，stone cutting machine 石材切割机
	section cutter，section cutting machine 型材切割机	section cutter，section cutting machine 型材切割机
	stripping machine 剥离机	stripping machine 剥离机
	angle polishing machine，angle grinder 角向磨光机	angle polishing machine，angle grinder 角向磨光机
	plain concrete cutter 混凝土切割机	plain concrete cutter 混凝土切割机
	concrete seam cutting machine 混凝土切缝机	concrete seam cutting machine 混凝土切缝机
	concrete drilling machine 混凝土钻孔机	concrete drilling machine 混凝土钻孔机
	terrazzo polishing machine，terrazzo grinder 水磨石磨光机	terrazzo polishing machine，terrazzo grinder 水磨石磨光机
	electric breaker 电镐	electric breaker 电镐

(续表)

Group/组	Type/型	Product/产品
other finishing machinery 其他装修机械	wallpaper machine 贴墙纸机	wallpaper machine 贴墙纸机
	spiral stone machine 螺旋洁石机	single spiral stone machine 单螺旋洁石机
	piercer 穿孔机	piercer 穿孔机
	duct grouting machine 孔道压浆机	duct grouting machine 孔道压浆机器
	pipe bending machine, tube bending machine 弯管机	pipe bending machine, tube bending machine 弯管机
	pipe threading cutting machine 管子套丝切断机	pipe threading cutting machine 管子套丝切断机
	pipe bending threading machine 管材弯曲套丝机	pipe bending threading machine 管材弯曲套丝机
	groove machine 坡口机	electric groove machine 电动坡口机
	elastic painting machine 弹涂机	electric elastic painting machine 电动弹涂机
	rolling painting machine 滚涂机	electric rolling painting machine 电动滚涂机

14 reinforced steel bar processing machinery and prestressed steel bar stretching machinery 钢筋及预应力机械

Group/组	Type/型	Product/产品
intensification machine for steel bar, steel bar reinforcement machinery 钢筋强化机械	steel bar cold-drawing machine, steel cold-drawing machine 钢筋冷拉机	winch steel bar cold-drawing machine, winch steel cold-drawing machine 卷扬机式钢筋冷拉机
		hydraulic steel bar cold-drawing machine, hydraulic steel cold-drawing machine 液压式钢筋冷拉机

（续表）

Group/组	Type/型	Product/产品
intensification machine for steel bar，steel bar reinforcement machinery 钢筋强化机械	steel bar cold-drawing machine，steel cold-drawing machine 钢筋冷拉机	roller steel bar cold-drawing machine，roller steel cold-drawing machine 滚轮式钢筋冷拉机
	steel drawbench 钢筋冷拔机	vertical steel drawbench 立式冷拔机
		horizontal steel drawbench 卧式冷拔机
		series steel drawbench 串联式冷拔机
	cold rolled ribbed bar forming machine，cold rolling steel wire and bar making machine 冷轧带肋钢筋成型机	active cold rolled ribbed bar forming machine，active cold rolling steel wire and bar making machine 主动冷轧带肋钢筋成型机
		passive cold rolled ribbed bar forming machine，passive cold rolling steel wire and bar making machine 被动冷轧带肋钢筋成型机
	cold-rolled-twisted bar forming machine，cold-rolled-twisted bar making machine 冷轧扭钢筋成型机	rectangular cold-rolled-twisted bar forming machine，rectangular cold-rolled-twisted bar making machine 长方形冷轧扭钢筋成型机
		square cold-rolled-twisted bar forming machine，square cold-rolled-twisted bar making machine 正方形冷轧扭钢筋成型机
	cold spirally-drewing deformed bar making machine，cold drawn spiral reinforcement making machine，cold spirally-drewing deformed bar forming machine，cold spirally-drewing deformed bar making machine 冷拔螺旋钢筋成型机	square cold spirally-drewing deformed bar making machine，square cold drawn spiral reinforcement making machine，square cold spirally-drewing deformed bar forming machine，square cold spirally-drewing deformed bar making machine 方形冷拔螺旋钢筋成型机
		round cold spirally-drewing deformed bar making machine，round cold drawn spiral reinforcement making machine，round cold spirally-drewing deformed bar forming machine，round cold spirally-drewing deformed bar making machine 圆形冷拔螺旋钢筋成型机

（续表）

Group/组	Type/型	Product/产品
steel bar processing machinery, steel processing machinery 单件钢筋成型机械	reinforcing bar cutting machine, steel bar cutting machine, steel bar shears 钢筋切断机	hand-held reinforcing bar cutting machine hand-held steel bar cutting machine hand-held steel bar shears 手持式钢筋切断机
		horizontal reinforcing bar cutting machine horizontal steel bar cutting machine horizontal steel bar shears 卧式钢筋切断机
		vertical reinforcing bar cutting machine vertical steel bar cutting machine vertical steel bar shears 立式钢筋切断机
		jaw reinforcing bar cutting machine jaw steel bar cutting machine jaw steel bar shears 颚剪式钢筋切断机
	reinforcing bar cutting line, reinforcing steel cutting line, steel bar cutting line 钢筋切断生产线	reinforcing bar shearing line reinforcing steel shearing line steel bar shearing line 钢筋剪切生产线
		reinforcing bar shearing line reinforcing steel shearing line steel bar shearing line 钢筋锯切生产线
	reinforcing bar straightening, cutting machine steel bar straightening, shearing machine 钢筋调直切断机	mechanical reinforcing bar straightening and cutting machine mechanical steel bar straightening and shearing machine 机械式钢筋调直切断机
		hydraulic reinforcing bar straightening and cutting machine hydraulic steel bar straightening and shearing machine 液压式钢筋调直切断机
		pneumatic reinforcing bar straightening and cutting machine pneumatic steel bar straightening and shearing machine 气动式钢筋调直切断机

（续表）

Group/组	Type/型	Product/产品
steel bar processing machinery, steel processing machinery 单件钢筋成型机械	angle bending machine, reinforcing bar bending machine, reinforcing steel bar bent machine 钢筋弯曲机	mechanical angle bending machine mechanical reinforcing bar bending machine mechanical reinforcing steel bar bent machine 机械式钢筋弯曲机
		hydraulic angle bending machine hydraulic reinforcing bar bending machine hydraulic reinforcing steel bar bent machine 液压式钢筋弯曲机
	angle bending line, reinforcing bar bending line, reinforcing steel bar bent line 钢筋弯曲生产线	vertical angle bending line vertical reinforcing bar bending line vertical reinforcing steel bar bent line 立式钢筋弯曲生产线
		horizontal angle bending line horizontal reinforcing bar bending line horizontal reinforcing steel bar bent line 卧式钢筋弯曲生产线
	spiral rebar bending machine, rebar spiral bending machine, steel bar spiral bender 钢筋弯弧机	mechanical spiral rebar bending machine mechanical rebar spiral bending machine mechanical steel bar spiral bender 机械式钢筋弯弧机
		hydraulic spiral rebar bending machine hydraulic rebar spiral bending machine hydraulic steel bar spiral bender 液压式钢筋弯弧机
	hoop bending machine for reinforcing bar, stirrup bender 钢筋弯箍机	CNC hoop bending machine for reinforcing bar CNC stirrup bender 数控钢筋弯箍机
	steel screw thread forming machine, steel thread making machine 钢筋螺纹成型机	steel taper thread forming machine steel taper thread making machine 钢筋锥螺纹成型机

(续表)

Group/组	Type/型	Product/产品
steel bar processing machinery, steel processing machinery 单件钢筋成型机械	steel screw thread forming machine, steel thread making machine 钢筋螺纹成型机	steel straight thread forming machine steel straight thread making machine steel parallel thread forming machine steel parallel thread making machine 钢筋直螺纹成型机
	steel screw thread forming line, steel thread making line 钢筋螺纹生产线	steel screw thread forming line steel thread making line 钢筋螺纹生产线
	reinforcement hoe machine 钢筋墩头机	reinforcement hoe machine 钢筋墩头机
combined steel bar processing machinery, combined steel processing machinery 组合钢筋成型机械	making machine for reinforcement mesh 钢筋网成型机	reinforcing steel mesh welding machine 钢筋网焊接成型机
	making machine for reinforcement cage 钢筋笼成型机	manual welding machine for reinforcement cage 手动焊接钢筋笼成型机
		automatic welding machine for reinforcement cage 自动焊接钢筋笼成型机
	making machine for reinforcement frame 钢筋桁架成型机	mechanical making machine for reinforcement frame 机械式钢筋桁架成型机
		hydraulic making machine for reinforcement frame 液压式钢筋桁架成型机
reinforcement connection machinery 钢筋连接机械	steel bar butt welder 钢筋对焊机	mechanical steel bar butt welder 机械式钢筋对焊机
		hydraulic steel bar butt welder 液压式钢筋对焊机
	concrete iron electro-slag pressure welding machine 钢筋电渣压力焊机	concrete iron electro-slag pressure welding machine 钢筋电渣压力焊机
	steel bar air pressure welding machine steel bar air pressure welder 钢筋气压焊机	closed steel bar air pressure welding machine closed steel bar air pressure welder 闭合式气压焊机

(续表)

Group/组	Type/型	Product/产品
reinforcement connection machinery 钢筋连接机械	steel bar air pressure welding machine, steel bar air pressure welder 钢筋气压焊机	open steel bar air pressure welding machine open steel bar air pressure welder 敞开式气压焊机
	sleeve squeezing machine for rebars, steel sleeve extruding machine 钢筋套筒挤压机	radial sleeve squeezing machine for rebars radial steel sleeve extruding machine 径向钢筋套筒挤压机
		axial sleeve squeezing machine for rebars axial steel sleeve extruding machine 轴向钢筋套筒挤压机
reinforced steel bar processing machinery and prestressed steel bar stretching machinery 预应力机械	prestressing steel wire heading machine 预应力钢筋墩头器	electric cold header 电动冷镦机
		hydraulic electric cold header 液压冷镦机
	prestressing steel tension jack 预应力钢筋张拉机	mechanical prestressing steel tension jack 机械式张拉机
		hydraulic prestressing steel tension jack 液压式张拉机
	prestressing steel tension penetrate strands machine 预应力钢筋穿束机	prestressing steel tension penetrate strands machine 预应力钢筋穿束机
		prestressing steel tension grouting machine 预应力钢筋灌浆机
	prestressing jack 预应力千斤顶	forward prestressing jack 前卡式预应力千斤顶
		continuous prestressing jack 连续式预应力千斤顶
prestressed machinery 预应力机具	prestressed anchorage device of reinforcement 预应力筋用锚具	forward prestressed anchorage device 前卡式预应力锚具
		center hole prestressed anchorage device 穿心式预应力锚具
	prestressed clamp of reinforcement 预应力筋用夹具	prestressed clamp of reinforcement 预应力筋用夹具

（续表）

Group/组	Type/型	Product/产品
prestressed machinery 预应力机具	prestressed connector of reinforcement 预应力筋用连接器	prestressed connector of reinforcement 预应力筋用连接器
other reinforced steel bar processing machinery and prestressed steel bar stretching machinery 其他钢筋及预应力机械		

15 rock drilling machine, rock drilling machinery 凿岩机械

Group/组	Type/型	Product/产品
rock drill, rock drilling machine 凿岩机	pneumatic hand-held rock drill 气动手持式凿岩机	hand-held rock drill 手持式凿岩机
	pneumatic rock drill 气动凿岩机	hand-held air-leg rock drill 手持气腿两用凿岩机
		air-leg rock drill 气腿式凿岩机
		air-leg high frequency rock drill 气腿式高频凿岩机
		pneumatic stoper 气动向上式凿岩机
		pneumatic drifter 气动导轨式凿岩机
		pneumatic drifter with independent rotation 气动导轨式独立回转凿岩机
	hand-held internal combustion rock drill 内燃手持式凿岩机	hand-held internal combustion rock drill 手持式内燃凿岩机
	hydraulic rock drill 液压凿岩机	hand-held hydraulic rock drill 手持式液压凿岩机

（续表）

Group/组	Type/型	Product/产品
rock drill，rock drilling machine 凿岩机	hydraulic rock drill 液压凿岩机	air-leg hydraulic rock drill 支腿式液压凿岩机
		hydraulic drifter 导轨式液压凿岩机
	electric rock drill 电动凿岩机	hand-held electric rock drill 手持式电动凿岩机
		air-leg electric rock drill 支腿式电动凿岩机
		electric drifter 导轨式电动凿岩机
open-air jumbo(drill) 露天钻车钻机	pneumatic and semi-hydraulic crawler open-air drill 气动、半液压履带式露天钻机	crawler open-air drill 履带式露天钻机
		crawler opencut downhole drill 履带式潜孔露天潜孔钻机
		crawler opencut medium pressure/high pressure downhole drill 履带式潜孔露天中压/高压潜孔钻机
	pneumatic and semi-hydraulic crawler open-air jumbo 气动、半液压轨轮式露天钻车	tire open-air jumbo 轮胎式露天钻车
		rail-mounted open-air jumbo 轨轮式露天钻车
	hydraulic crawler drill 液压履带式钻机	crawler open-air hydraulic drill 履带式露天液压钻机
		crawler opencut hydraulic downhole drill 履带式露天液压潜孔钻机
	hydraulic jumbo 液压钻车	tire open-air hydraulic jumbo 轮胎式露天液压钻车
		rail-mounted open-air hydraulic jumbo 轨轮式露天液压钻车
underground jumbo(drill) 井下钻车钻机	pneumatic and semi-hydraulic crawler drill 气动、半液压履带式钻机	crawler mining drill 履带式采矿机
		crawler tunnelling drill 履带式掘进钻机
		crawler bolting rotary drill 履带式锚杆钻机

(续表)

Group/组	Type/型	Product/产品
underground jumbo(drill) 井下钻车钻机	pneumatic and semi-hydraulic crawler jumbo 气动、半液压式钻车	tire mining/tunnelling/bolt jumbo 轮胎式采矿/掘进/锚杆钻车
		rail-mounted mining/tunnelling/bolt jumbo 轨轮式采矿/掘进/锚杆钻车
	full-hydraulic crawler drill 全液压履带式钻机	crawler hydraulic mining/tunnelling/bolt drill 履带式液压采矿/掘进/锚杆钻机
	full-hydraulic crawler jumbo 全液压钻车	tire hydraulic mining/tunnelling/bolt jumbo 轮胎式液压采矿/掘进/锚杆钻车
		rail-mounted hydraulic mining/tunnelling/bolt jumbo 轨轮式液压采矿/掘进/锚杆钻车
pneumatic downhole drill hammer 气动潜孔冲击器	low pressure downhole drill hammer 低气压潜孔冲击器	downhole drill hammer 潜孔冲击器
	medium pressure/high pressure downhole drill hammer 中、高气压潜孔冲击器	medium pressure/high pressure downhole drill hammer 中压/高压潜孔冲击器
rock drilling auxiliary 凿岩辅助设备	leg, leg support, outrigger, stabilizer 支腿	air leg/water leg/oil leg/hand cranking leg 气腿/水腿/油腿/手摇式支腿
	pillar rig 柱式钻架	single-column drill rig double-column drill rig 单柱式/双柱式钻架
	ring guide drill rig 圆盘式钻架	ring guide /shaft/ring drill rig 圆盘式/伞式/环形钻架
	others 其他	dust collector ring lineoiler bit grinder 集尘器、注油器、磨钎机
other rock drilling machine 其他凿岩机械		

16 air tool, pneumatic tool 气动工具

Group/组	Type/型	Product/产品
rotary pneumatic tool 回转式气动工具	engraving pen 雕刻笔	pneumatic engraving pen 气动雕刻笔
	pneumatic drill 气钻	straight pneumatic drill pistol-grip pneumatic drill side-handle pneumatic drill combined pneumatic drill pneumatic craniotomy drill pneumatic dental drill 直柄式/枪柄式/侧柄式/组合用气钻/气动开颅钻/气动牙钻
	tapping machine 攻丝机	straight pneumatic tapping machine pistol-grip pneumatic tapping machine combined pneumatic tapping machine 直柄式/枪柄式/组合用气动攻丝机
	grinder 砂轮机	straight pneumatic grinder angular pneumatic grinder vertical pneumatic grinder combined pneumatic grinder straight pneumatic wire brush 直柄式/角向/端面式/组合气动砂轮机/直柄式气动钢丝刷
	polisher 抛光机	vertical polisher circumferential polisher angular polisher 端面/圆周/角向抛光机
	refacer 磨光机	vertical pneumatic refacer circumferential pneumatic refacer reciprocating pneumatic refacer sand belt pneumatic refacer skateboard pneumatic refacer triangular pneumatic refacer 端面/圆周/往复式/砂带式/滑板式/三角式气动磨光机
	milling cutter 铣刀	air milling cutter angular air milling cutter 气铣刀/角式气铣刀
	pneumatic shear 气锯	band pneumatic shear belt swing pneumatic shear disc pneumatic shear chain pneumatic shear 带式/带式摆动/圆盘式/链式气锯

(续表)

Group/组	Type/型	Product/产品
rotary pneumatic tool 回转式气动工具	pneumatic shear 气锯	pneumatic saw 气动细锯
	shear 剪刀	pneumatic shear pneumatic punching shear 气动剪切机/气动冲剪机
	air screw 气螺刀	straight stall air screw pistol-grip stall air screw angular stall air screw 直柄式/枪柄式/角式失速型气螺刀
	pneumatic wrench 气扳机	pistol-grip stall non-impact pneumatic wrench clutch non-impact pneumatic wrench automatic closed non-impact pneumatic wrench 枪柄式失速型/离合型/自动关闭型纯扭气扳机
		angular stall non-impact pneumatic wrench clutch non-impact pneumatic wrench 角式失速型/离合型纯扭气扳机
		ratchet non-impact pneumatic wrench two-speed non-impact pneumatic wrench combination non-impact pneumatic wrench 棘轮式/双速型/组合式全扭气扳机
		open claw sleeve non-impact pneumatic wrench close non-impact pneumatic wrench 开口爪型套筒/闭口爪型套筒纯扭气扳机
		pneumatic stud pneumatic wrench 气动螺柱气扳机
		straight pneumatic wrench straight torque controlled pneumatic wrench 直柄式/直柄式定扭矩气扳机
		energy storage pneumatic wrench 储能型气扳机
		straight high-speed pneumatic wrench 直柄式高速气扳机

（续表）

Group/组	Type/型	Product/产品
rotary pneumatic tool 回转式气动工具	pneumatic wrench 气扳机	pistol-grip pneumatic wrench pistol-grip torque controlled pneumatic wrench pistol-grip high-speed pneumatic wrench 枪柄式/枪柄式定扭矩/枪柄式高速气扳机
		angular pneumatic wrench angular torque controlled pneumatic wrench angular high-speed pneumatic wrench 角式/角式定扭矩/角式高速气扳机
		combined pneumatic wrench 组合式气扳机
		straight pneumatic pulse wrench pistol-grip pneumatic pulse wrench angular pneumatic pulse wrench electrically controlled pneumatic pulse wrench 直柄式/枪柄式/角式/电控型脉冲气扳机
	vibrator 振动器	rotary pneumatic vibrator 回转式气动振动器
percussive pneumatic tool, pneumatic impact tools 冲击式气动工具	riveting hammer 铆钉机	straight pneumatic riveting hammer curved-handle pneumatic riveting hammer pistol-grip pneumatic riveting hammer 直柄式/弯柄式/枪柄式气动铆钉机
		pneumatic pull-type rivetter pneumatic push-type rivetter 气动拉铆钉机/压铆钉机
	nailing machine 打钉机	air nailer strip nail air nailer U-type nail air nailer 气动打钉机/条形钉/U 型钉气动打钉机
	stapler 订合机	pneumatic stapler 气动订合机
	press brake 折弯机	press brake 折弯机

(续表)

Group/组	Type/型	Product/产品
percussive pneumatic tool, pneumatic impact tools 冲击式气动工具	printer 打印器	printer 打印器
	pliers 钳	pneumatic pliers hydraulic pliers 气动钳/液压钳
	splitter 劈裂机	pneumatic splitter hydraulic splitter 气动/液压劈裂机
	expander, dilator 扩张器	hydraulic expander 液压扩张机
	hydraulic cutter 液压剪	hydraulic cutter 液压剪
	mixer 搅拌机	pneumatic mixer, pneumatic stirrer 气动搅拌机
	strapping machine 捆扎机	pneumatic strapping machine 气动捆扎机
	sealing machine 封口机	pneumatic sealing machine 气动封口机
	breaking hammer 破碎锤	pneumatic breaking hammer 气动破碎锤
	pickaxe 镐	pneumatic pickaxe hydraulic pickaxe internal combustion pickaxe electrical pickaxe 气镐、液压镐、内燃镐、电动镐
	pneumatic chipping hammer 气铲	straight pneumatic chipping hammer curved-handle pneumatic chipping hammer annular-handle pneumatic chipping hammer shovel machine 直柄式/弯柄式/环柄式气铲/铲石机
	tamping machine 捣固机	pneumatic tamper sleeper tamper earth tamper 气动捣固机/枕木捣固机/夯土捣固机
	file 锉刀	rotary pneumatic file reciprocating pneumatic file rotary reciprocating air file rotary swing pneumatic file 旋转式/往复式/旋转往复式/旋转摆动式气锉刀

(续表)

Group/组	Type/型	Product/产品
percussive pneumatic tool, pneumatic impact tools 冲击式气动工具	blade 刮刀	pneumatic blade pneumatic swing blade 气动刮刀/气动摆动式刮刀
	engraving tool 雕刻机	percussive pneumatic engraving tool 回转式气动雕刻机
	scrabble 凿毛机	pneumatic scrabble 气动凿毛机
	vibrator 振动器	pneumatic vibrating rod 气动振动棒
		percussive vibrator 冲击式振动器
other pneumatic machinery 其他气动机械	pneumatic motor 气动马达	pneumatic vane motor, vane pneumatic motor 叶片式气动马达
		piston pneumatic motor axial piston type pneumatic motor 活塞式/轴向活塞式气动马达
		gear pneumatic motor, pneumatic gear motor 齿轮式气动马达
		pneumatic turbine motor, turbo-pneumatic motor 透平式气动马达
	pneumatic pump 气动泵	pneumatic pump 气动泵
		pneumatic diaphragm pump 气动隔膜泵
	air hoist, pneumatic hoist 气动吊	loop-chain type pneumatic hoist wire rope pneumatic hoist 环链式/钢绳式气动吊
	pneumatic hoist 气动绞车/绞盘	pneumatic hoist 气动绞车
	pneumatic pile driver, pneumatic pile hammer 气动桩机	pneumatic pile driver, pneumatic pile drawer 气动打桩机/拔桩机
other pneumatic tools 其他气动工具		

17 military construction machinery 军用工程机械

Group/组	Type/型	Product/产品
road machinery 道路机械	armoured construction vehicle 装甲工程车	crawler armoured construction vehicle 履带式装甲工程车
		wheeled armoured construction vehicle 轮式装甲工程车
	mutiple purpose construction vehicle 多用工程车	crawler mutiple purpose construction vehicle 履带式多用工程车
		wheeled mutiple purpose construction vehicle 轮式多用工程车
	bulldozer 推土机	crawler bulldozer 履带式推土机
		wheeled bulldozer 轮式推土机
	loader 装载机	wheeled loader 轮式装载机
		skid loader 滑移装载机
	grader, plough, plow 平地机	self-propelled grader 自行式平地机
	road roller 压路机	vibratory road roller 振动式压路机
		static road roller 静作用式压路机
	snow removal truck 除雪机	rotor snow removal truck 轮子式除雪机
		snow plough 犁式除雪机
field city building machinery 野战筑城机械	trencher 挖壕机	crawler trencher 履带式挖壕机
		wheeled trencher 轮式挖壕机
	digging machine 挖坑机	crawler digging machine 履带式挖坑机
		wheeled digging machine 轮式挖坑机

（续表）

Group/组	Type/型	Product/产品
field city building machinery 野战筑城机械	excavator 挖掘机	crawler excavator 履带式挖掘机
		wheel excavator 轮式挖掘机
		mountain excavator 山地挖掘机
	field work machinery 野战工事作业机械	field work vehicle 野战工事作业车
		mountain jungle work machine 山地丛林作业机
	drilling machine 钻孔机具	soil drill 土钻
		fast drilling rig 快速成孔钻机
	frozen soil working machinery 冻土作业机械	mechanical dynamite trencher 机-爆式挖壕机
		frozen soil drilling machine 冻土钻井机
permanent city building machinery 永备筑城机械	rock drill，rock drilling machine 凿岩机	rock drill，rock drilling machine 凿岩机
		drilling jumbo 凿岩台车
	air compressor 空压机	motor air compressor 电动机式空压机
		internal combustion engine air compressor 内燃机式空压机
	tunnel ventilator 坑道通风机	tunnel ventilator 坑道通风机
	tunnel joint boring machine 坑道联合掘进机	tunnel joint boring machine 坑道联合掘进机
	tunnel rock loader 坑道装岩机	tunnel rock loader 坑道式装岩机
		tire rock loader 轮胎式装岩机

（续表）

Group/组	Type/型	Product/产品
permanent city building machinery 永备筑城机械	tunnel covering machinery 坑道被覆机械	steel trolley 钢模台车
		concrete casting machine 混凝土浇注机
		screw concrete spraying machine, spiral concrete sprayer 混凝土喷射机
	crusher, grinder 碎石机	jaw crusher 颚式碎石机
		conical crusher 圆锥式碎石机
		roll type crusher 辊式碎石机
		hammer mill, hammer crusher 锤式碎石机
	screening machine 筛分机	drum type screening machine 滚筒式筛分机
	concrete mixer, truck mixer agitator 混凝土搅拌机	inverted concrete mixer 倒翻式凝土搅拌机
		inclined concrete mixer 倾斜式凝土搅拌机
		rotary concrete mixer 回转式凝土搅拌机
	steel bar processing machinery, steel processing machinery 钢筋加工机械	straight rib-shearing machine 直筋-切筋机
		bending machine 弯筋机
	wood processing machinery 木材加工机械	motorcycle saw 摩托锯
		circular saw 圆锯机
mine laying, detection and sweeping machinery 布、探、扫雷机械	mine laying machine 布雷机械	crawler mine laying vehicle 履带式布雷车
		wheel mine laying vehicle 轮胎式布雷车
	mine detection machinery 探雷机械	road mine detection vehicle 道路探雷车

（续表）

Group/组	Type/型	Product/产品
mine laying, detection and sweeping machinery 布、探、扫雷机械	minesweeper 扫雷机械	mechanical minesweeper 机械式扫雷车
		integrated minesweeper 综合式扫雷车
bridge beam erecting machine, bridge girder-erecting machinery 架桥机械	bridge erection machinery 架桥作业机械	bridge-erecting vehicle 架桥作业车
	mechanized bridge 机械化桥	crawler mechanized bridge 履带式机械化桥
		tire mechanized bridge 轮胎式机械化桥
	pile driving machinery 打桩机械	pile driver 打桩机
water supply machinery 野战给水机械	water source reconnaissance vehicle 水源侦察车	water source reconnaissance vehicle 水源侦察车
	miser 钻井机	rotary miser 回转式钻井机
		impact miser 冲击式钻井机
	dainage machinery 汲水机械	internal combustion water pump 内燃抽水机
		electric water pump 电动抽水机
	water purification machinery 净水机械	self-propelled water purifier 自行式净水车
		towed water purifier 拖式净水车
camouflage mechanism 伪装机械	camouflage survey vehicle 伪装勘测车	camouflage survey vehicle 伪装勘测车
	camouflage vehicle 伪装作业车	pattern painting working vehicle 迷彩作业车
		fake target production car 假目标制作车
		barrier (high altitude) operating vehicle 遮障(高空)作业车

(续表)

Group/组	Type/型	Product/产品
safeguard operation vehicles 保障作业车辆	mobile power station 移动式电站	self-propelled mobile power station 自行式移动式电站
		towed mobile power station 拖式移动式电站
	metal and wood engineering work vehicle 金木工程作业车	metal and wood engineering work vehicle 金木工程作业车
	hoisting machinery 起重机械	mobile crane, truck crane, truck mounted crane 汽车起重机
		rubber tyred crane, tyre crane, tyre-mounted crane, wheel crane 轮胎式起重机
	hydraulic maintenance vehicle 液压检修车	hydraulic maintenance vehicle 液压检修车
	construction machinery repair vehicle 工程机械修理车	construction machinery repair vehicle 工程机械修理车
	special tractor 专用牵引车	special tractor 专用牵引车
	source truck 电源车	source truck 电源车
	air supply car 气源车	air supply car 气源车
other military construction machinery 其他军用工程机械		

18 lift and escalator 电梯及扶梯

Group/组	Type/型	Product/产品
lift 电梯	passenger lift 乘客电梯	alternating current passenger lift 交流乘客电梯

(续表)

Group/组	Type/型	Product/产品
lift 电梯	passenger lift 乘客电梯	direct current passenger lift 直流乘客电梯
		hydraulic current passenger lift 液压乘客电梯
	freight lift 载货电梯	alternating current freight lift 交流载货电梯
		hydraulic current freight lift 液压载货电梯
	passenger-goods lift 客货电梯	alternating current passenger-goods lift 交流客货电梯
		direct current passenger-goods lift 直流客货电梯
		hydraulic current passenger-goods lift 液压客货电梯
	bed lift 病床电梯	alternating current bed lift 交流病床电梯
		hydraulic bed lift 液压病床电梯
	residential lift 住宅电梯	alternating current residential lift 交流住宅电梯
	dumbwaiter, dumbwaiter lift, service lift 杂物电梯	alternating current dumbwaiter lift 交流杂物电梯
	observation lift, panoramic lift 观光电梯	alternating current observation lift 交流观光电梯
		current current observation lift 直流观光电梯
		hydraulic current observation lift 液压观光电梯
	lift on ships 船用电梯	alternating current lift on ships 交流船用电梯
		hydraulic lift on ships 液压船用电梯
	vehicle lift 车辆用电梯	alternating current vehicle lift 交流车辆用电梯
		hydraulic vehicle lift 液压车辆用电梯

（续表）

Group/组	Type/型	Product/产品
lift 电梯	blast defense lift 防爆电梯	blast defense lift 防爆电梯
auto-escalator 自动扶梯	common auto-escalator 普通型自动扶梯	common chain auto-escalator 普通型链条式自动扶梯
		common rack auto-escalator 普通型齿条式自动扶梯
	public auto-escalator 公共交通型自动扶梯	public chain auto-escalator 公共交通型链条式自动扶梯
		public rack auto-escalator 公共交通型齿条式自动扶梯
	spiral auto-escalator 螺旋形自动扶梯	spiral auto-escalator 螺旋形自动扶梯
moving sidewalk, moving walkway, passenger conveyer 自动人行道	common moving walkway 普通型自动人行道	common pedal type moving walkway 普通型踏板式自动人行道
		common belt drum type moving walkway 普通型胶带滚筒式自动人行道
	public transport type moving walkway 公共交通型自动人行道	public transport pedal type moving walkway 公共交通型踏板式自动人行道
		public transport belt drum type moving walkway 公共交通型胶带滚筒式自动人行道
other lifts and escalators 其他电梯及扶梯		

19 construction machinery parts 工程机械配套件

Group/组	Type/型	Product/产品
power system 动力系统	diesel, internal combustion engine, gas engine 内燃机	diesel engine 柴油发动机
		gasoline engine 汽油发动机
		gas engine 燃气发动机
		dual-power engine 双动力发动机

（续表）

Group/组	Type/型	Product/产品
power system 动力系统	power accumulator battery 动力蓄电池组	power accumulator battery 动力蓄电池组
	accessory device 附属装置	water radiator（water tank）水散热箱（水箱）
		oil cooler 机油冷却器
		cooling fan 冷却风扇
		fuel tank 燃油箱
		turbo-supercharger，turbo-charger 涡轮增压器
		air cleaner，air filter，air strainer 空气滤清器
		oil filter 机油滤清器
		diesel filter 柴油滤清器
		exhaust pipe（muffler）assembly 排气管（消声器）总成
		air compressor 空气压缩机
		generator 发电机
		starter-motor 启动马达
transmission system 传动系统	clutch，clutch mechanism，positive clutch 离合器	dry clutch 干式离合器
		wet clutch 湿式离合器
	torque converter 变矩器	hydraulic torque converter 液力变矩器
		hydraulic coupling unit 液力耦合器
	gearbox，transmission 变速器	mechanical transmission 机械式变速器

（续表）

Group/组	Type/型	Product/产品
transmission system 传动系统	gearbox, transmission 变速器	power transmission 动力换挡变速器
		electric-hydraulic transmission 电液换挡变速器
	driving motor 驱动电机	DC motor 直流电机
		AC motor 交流电机
	drive shaft device 传动轴装置	drive shaft, driving shaft, propeller shaft, transmission shaft, weight drive shaft to motor 传动轴
		coupler 联轴器
	driving axle, driving shaft 驱动桥	driving axle, driving shaft 驱动桥
	decelerator 减速器	final drive 终传动
		wheel reductor 轮边减速
hydraulic packing bush 液压密封装置	cylinder 油缸	medium and low pressure cylinder 中低压油缸
		high pressure cylinder 高压油缸
		ultra high pressure cylinder 超高压油缸
	hydraulic pump, hydraulic system pump 液压泵	gear pump 齿轮泵
		vane pump 叶片泵
		piston pump 柱塞泵
	fluid motor, hydraulic motor, hydraulic system motor, hydromotor 液压马达	gear motor 齿轮马达
		vane motor 叶片马达
		piston motor 柱塞马达

(续表)

Group/组	Type/型	Product/产品
hydraulic packing bush 液压密封装置	hydraulic valve 液压阀	hydraulic multi-way directional valves 液压多路换向阀
		pressure control valve，pressure-regulating valve 压力控制阀
		flow control valve 流量控制阀
		hydraulic pilot valve 液压先导阀
	hydraulic reducer 液压减速机	traveling reducer 行走减速机
		rotary reducer 回转减速机
	accumulator 蓄能器	accumulator 蓄能器
	central swing tower 中央回转体	central swing tower 中央回转体
	hydraulic pipe 液压管件	high pressure hose 高压软管
		low pressure hose 低压软管
		high temperature and low pressure hose 高温低压软管
		hydraulic metal connecting pipe 液压金属连接管
		hydraulic pipe fitting 液压管接头
	hydraulic system accessories 液压系统附件	hydraulic oil filter 液压油滤油器
		hydraulic oil cooler 液压油散热器
		hydraulic oil tank 液压油箱
	gland，packing bush 密封装置	moving oil seal 动油封件
		fixed seal 固定密封件

(续表)

Group/组	Type/型	Product/产品
brake system 制动系统	air reservoir 贮气筒	air reservoir 贮气筒
	pneumatic valve 气动阀	pneumatic reversing valve 气动换向阀
		pneumatic pressure control valve 气动压力控制阀
	intensifier pump assembly 加力泵总成	intensifier pump assembly 加力泵总成
	pneumatic brake fitting 气制动管件	pneumatic hose 气动软管
		pneumatic metal pipe 气动金属管
		pneumatic pipe fitting 气动管接头
	lubricant distributor, oil separator 油水分离器	lubricant distributor, oil separator 油水分离器
	brake pump 制动泵	brake pump 制动泵
	brake 制动器	parking brake 驻车制动器
		disc brake 盘式制动器
		band brake 带式制动器
		wet disc brake 湿式盘式制动器
walking device 行走装置	tyre assembly 轮胎总成	solid tyre 实心轮胎
		pneumatic tyre 充气轮胎
	rim assembly, wheel rim assembly 轮辋总成	rim assembly, wheel rim assembly 轮辋总成
	tyre non-skid chain 轮胎防滑链	tyre non-skid chain 轮胎防滑链

（续表）

Group/组	Type/型	Product/产品
walking device 行走装置	track assembly 履带总成	ordinary track assembly 普通履带总成
		wet track assembly 湿式履带总成
		rubber track assembly 橡胶履带总成
		triple track assembly 三联履带总成
	four wheels 四轮	bearing wheel assembly 支重轮总成
		carrier roller assembly 拖链轮总成
		track idler assembly 引导轮总成
		drive sprocket assembly 驱动轮总成
	track tensioner assembly 履带张紧装置总成	track tensioner assembly 履带张紧装置总成
guidance system 转向系统	commutator assembly 转向器总成	commutator assembly 转向器总成
	steering axle 转向桥	steering axle 转向桥
	steering device 转向操作装置	reversing gear，steering apparatus，steering gear 转向装置
vehicle frame and working device 车架及工作装置	center-articulated frame，chassis，frame，vehicle frame 车架	center-articulated frame，chassis，frame，vehicle frame 车架
		slewing ring 回转支撑
		driver’s cab，driving cab，operator cab 驾驶室
		driver’s seat assembly 司机座椅总成
	working device 工作装置	boom 动臂

（续表）

Group/组	Type/型	Product/产品
vehicle frame and working device 车架及工作装置	working device 工作装置	bucket rod 斗杆
		bucket 铲/挖斗
		bucket tooth 斗齿
		blade，cutting edge 刀片
	balance weight，counterweight 配重	balance weight，counterweight 配重
	door frame system 门架系统	door frame 门架
		chain 链条
		fork，fork arm，pallet fork，shuttle 货叉
	hoisting equipment 吊装装置	hook 吊钩
		jib 臂架
	vibrating device 振动装置	vibrating device 振动装置
electrical device 电器装置	electronic control system assembly 电控系统总成	electronic control system assembly 电控系统总成
	composite instrument assembly 组合仪表总成	composite instrument assembly 组合仪表总成
	monitor assembly 监控器总成	monitor assembly 监控器总成
	instrument 仪表	time counter 计时表
		speed counter，tachometer 速度表
		thermometer 温度表

(续表)

Group/组	Type/型	Product/产品
electrical device 电器装置	instrument 仪表	oil pressure gauge 油压表
		barometer 气压表
		oil level gauge 油位表
		ampere meter 电流表
		voltmeter 电压表
	alarm 报警器	traffic alarm 行车报警器
		backing car annunciator 倒车报警器
	car light 车灯	lighting lamp 照明灯
		turn signal lamp 转向指示灯
		brake indicator 刹车指示灯
		fog lamp 雾灯
		driver's cab top light 司机室顶灯
	air conditioner 空调器	air conditioner 空调器
	unit heater 暖风机	unit heater 暖风机
	electric fan 电风扇	electric fan 电风扇
	window cleaner，wiper 刮水器	window cleaner，wiper 刮水器
	accumulator 蓄电池	accumulator 蓄电池

（续表）

Group/组	Type/型	Product/产品
special fittings 专用属具	hydraulic hammer，hydraulically powered impact hammer 液压锤	hydraulic hammer，hydraulically powered impact hammer 液压锤
	hydraulic shear 液压剪	hydraulic shear 液压剪
	hydraulic vice 液压钳	hydraulic vice 液压钳
	ripper 松土器	ripper 松土器
	wooden fork 夹木叉	wooden fork 夹木叉
	forklift truck special fittings 叉车专用属具	forklift truck special fittings 叉车专用属具
	other fittings 其他属具	other fittings 其他属具
other accessories 其他配套件		

20 other special construction machinery 其他专用工程机械

Group/组	Type/型	Product/产品
special construction machinery for power station 电站专用工程机械	self-pulling tower crane 扳起式塔式起重机	self-pulling tower crane for power station 电站专用扳起式塔式起重机
	self-raising tower crane 自升式塔式起重机	self-raising tower crane for power station 电站专用自升塔式起重机
	furnace roof crane 锅炉炉顶起重机	furnace roof crane for power station 电站专用锅炉炉顶起重机
	portal crane，portal jib crane，portal slewing crane 门座起重机	portal crane for power station 电站专用门座起重机
	crawler crane 履带式起重机	crawler crane for power station 电站专用履带式起重机

(续表)

Group/组	Type/型	Product/产品
special construction machinery for power station 电站专用工程机械	gantry crane 龙门式起重机	gantry crane for power station 电站专用龙门式起重机
	cable type crane 缆索起重机	translational elevated cable crane for power station 电站专用平移式高架缆索起重机
	cargo hoist, hoisting plant, lifter, lifting, lifting apparatus 提升装置	cables hydraulic lifting device for power station 电站专用钢索液压提升装置
	construction hoist 施工升降机	construction hoist for power station 电站专用施工升降机
		curve construction elevator 曲线施工电梯
	concrete mixing plant 混凝土搅拌楼	concrete mixing plant for power station 电站专用混凝土搅拌楼
	concrete batch plant 混凝土搅拌站	concrete batch plant for power station 电站专用混凝土搅拌站
	tower belt 塔带机	tower belt distributor 塔式皮带布料机
construction and maintenance machinery of rail transit 轨道交通施工与养护工程机械	bridge beam erecting machine 架桥机	bridge erecting machine for concrete box girder of high-speed passenger dedicated line 高速客运专线混凝土箱梁架桥机
		bridge erecting machine for concrete box girder without guide beam for high-speed passenger dedicated line 高速客运专线无导梁式混凝土箱梁架桥机
		bridge erecting machine for concrete box girder with guide beam for high-speed passenger dedicated line 高速客运专线导梁式混凝土箱梁架桥机
		bridge erecting machine for concrete box girder with low guide beam for high-speed passenger dedicated line 高速客运专线下导梁式混凝土箱梁架桥机

(续表)

Group/组	Type/型	Product/产品
construction and maintenance machinery of rail transit 轨道交通施工与养护工程机械	bridge beam erecting machine 架桥机	wheel-rail moving concrete box girder bridge erecting machine for high-speed passenger dedicated line 高速客运专线轮轨走行移位式混凝土箱梁架桥机
		wheel-rail moving concrete box girder bridge erecting machine with solid rubber wheel 实胶轮走行移位式混凝土箱梁架桥机
		mixed wheel-rail moving concrete box girder bridge erecting machine 混合走行移位式混凝土箱梁架桥机
		double-track box girder bridge erector for high-speed passenger dedicated line passing through tunnel 高速客运专线双线箱梁过隧道架桥机
		conventional railway T-beam bridge machine 普通铁路 T 梁架桥机
		conventional highway-railway T-beam bridge machine 普通铁路公铁两用 T 梁架桥机
	girder carrier 运梁车	tire type girder car for double-track concrete box girder on high speed passenger dedicated line 高速客运专线混凝土箱梁双线箱梁轮胎式运梁车
		tire type girder car for double-track concrete box girder on high speed passenger dedicated line passing through tunnel 高速客运专线过隧道双线箱梁轮胎式运梁车
		tire type girder car for single-track box girder on high speed passenger dedicated line 高速客运专线单线箱梁轮胎式运梁车
		conventional railway rail-mounted T-beam transporter 普通铁路轨行式 T 梁运梁车

(续表)

Group/组	Type/型	Product/产品
construction and maintenance machinery of rail transit 轨道交通施工与养护工程机械	girder lifter for beam yard 梁场用提梁机	tire girder lifter 轮胎式提梁机
		wheel-rail girder lifter 轮轨式提梁机
	track superstructure manufacturing, transportation and paving equipment 轨道上部结构制运铺设备	long rail single sleeper transportation and paving equipment for ballasted lines 有砟线路长轨单枕法运铺设备
		manufacturing, transportation and paving equipment for ballastless track system 无砟轨道系统制运铺设备
		manufacturing, transportation and paving equipment for ballastless slab track system 无砟板式轨道系统制运铺设备
		manufacturing, transportation and paving equipment for ballastless track system 无砟轨道系统制运铺设备
		manufacturing, transportation and paving equipment for ballastless slab track system 无砟板式轨道系统制运铺设备
	ballast maintenance equipment series 道砟设备养护用设备系列	special ballast truck 专用运道砟车
		ballast shaping machine 配砟整形机
		ballast stamping machine 道砟捣固机
		ballast cleaning machine 道砟清筛机
	construction and maintenance equipment of electrified circuit 电气化线路施工与养护设备	contact system catenary pillar digger 接触网立柱挖坑机
		contact system catenary column erecting equipment 接触网立柱竖立设备
		contact system catenary laying car 接触网架线车

(续表)

Group/组	Type/型	Product/产品
special construction machinery for water conservancy 水利专用工程机械	special construction machinery for water conservancy 水利专用工程机械	special construction machinery for water conservancy 水利专用工程机械
special construction machinery for mines 矿山用工程机械	special construction machinery for mines 矿山用工程机械	special construction machinery for mines 矿山用工程机械
other construction machinery 其他工程机械